PHYSICS IN FOCUS

SKILLS AND ASSESSMENT
WORKBOOK

YEAR 12

Adam Sloan
Edward Baker
Darren Goossens
Owen Hamerton

NELSON
A Cengage Company

Physics in Focus: Skills and Assessment Workbook Year 12
1st Edition
Adam Sloan
Edward Baker
Darren Goossens
Owen Hamerton
ISBN 9780170449687

Publisher: Sam Bonwick
Editorial manager: Simon Tomlin
Editor: Elaine Cochrane
Proofreader: Marta Veroni
Original cover design by Chris Starr (MakeWork), Adpated by: Justin Lim
Text design: Ruth Comey (Flint Design)
Project designer: Justin Lim
Permissions researcher: Wendy Duncan
Production controller: Alice Kane
Typeset by: MPS Limited

Any URLs contained in this publication were checked for currency during the production process. Note, however, that the publisher cannot vouch for the ongoing currency of URLs.

Acknowledgements

Cover image: iStock.com/Brostock.
Inquiry questions on pages 4, 11, 21, 37, 46, 50, 57, 68, 77, 86, 94, 106, 120, 127, 138 and 149 are from the Physics Stage 6 Syllabus © NSW Education Standards Authority for and on behalf of the Crown in right of the State of New South Wales, 2017.

For product information and technology assistance,
in Australia call **1300 790 853**;
in New Zealand call **0800 449 725**

For permission to use material from this text or product, please email
aust.permissions@cengage.com

ISBN 978 0 17 044968 7

Cengage Learning Australia
Level 7, 80 Dorcas Street
South Melbourne, Victoria Australia 3205

Cengage Learning New Zealand
Unit 4B Rosedale Office Park
331 Rosedale Road, Albany, North Shore 0632, NZ

For learning solutions, visit **cengage.com.au**

Printed in China by 1010 Printing International Limited.
1 2 3 4 5 6 7 25 24 23 22 21

CONTENTS

MODULE EIGHT » FROM THE UNIVERSE TO THE ATOM   103

9780170449687

ABOUT THIS BOOK

FEATURES

▶ Reviewing prior knowledge from Year 11 starts each module and Checking understanding of key concepts ends each module

▶ Learning goals stated at the top of each worksheet set the intention and help you understand what is required

▶ Chapters clearly follow the sequence of the syllabus and are organised by inquiry question

▶ Page references to the content-rich *Physics in Focus Year 12* student book provide an integrated learning experience

▶ Brief content summaries are provided where applicable

▶ Hint boxes provide guidance on how to answer questions effectively

▶ Complete practice exam is provided

▶ Fully worked solutions appear at the back of the book to allow you to work independently and check your progress

ORGANISATION OF YOUR WORKBOOK

Each chapter begins with the relevant inquiry question and follows the sequence of the syllabus. Worksheets have been designed to complement the student book and provide additional opportunities to apply and revise your learning. Completion of these worksheets will provide you with a solid foundation to complete assessments and depth studies.

The workbook ends with a practice exam for you to complete independently. Use the fully worked solutions and marking criteria in the answers section to self-evaluate and plan for improvement.

Note: the formulae and data used in this workbook are consistent with those used in the HSC exam. Formulae and data sheets are available to students and teachers on the Physics Stage 6 Syllabus page of the NESA website.

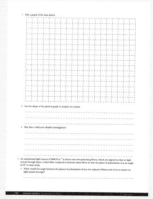

State numerical answers correct to an appropriate number of significant figures.

Reviewing prior knowledge

1 Average speed $= \dfrac{\text{Total distance}}{\text{Total time}}$ is an equation that can be used for all moving objects.

Use the equation to find:

a the distance travelled, in metres, by a car driving at $60\,\text{km}\,\text{h}^{-1}$ for 1.5 hours.

_____

b the time taken, in seconds, by a jogger running at $5.0\,\text{m}\,\text{s}^{-1}$ to travel $0.40\,\text{km}$.

_____

c the average speed of a cyclist, in $\text{m}\,\text{s}^{-1}$, who travels $25.0\,\text{km}$ in 66 minutes and 40 seconds.

_____

2 **a** Define acceleration.

b Use your definition to decide whether the moving object in the following situations undergoes acceleration. Justify your choice.

i A rubber ball approaches a wall at $10\,\text{m}\,\text{s}^{-1}$ and bounces off at the same speed.

_____

ii A train travels along a straight track at a constant speed.

_____

iii A car enters a corner on the road and leaves the corner at the same speed.

_____

iv A car rolls down a steep hill without the brakes being applied.

_____

3 The three equations below are referred to as the equations of motion and are applicable in situations of constant acceleration.

$v = u + at$

$v^2 = u^2 + 2as$

$s = ut + \dfrac{1}{2}at^2$

a Rearrange $v = u + at$ to make a the subject.

_____

b Rearrange $v^2 = u^2 + 2as$ to make u the subject.

_____

c Rearrange $s = ut + \dfrac{1}{2}at^2$ to make t the subject, assuming $u = 0$.

_____

4 A spacecraft descends directly downwards towards the Moon's surface at $60\,\text{m s}^{-1}$. To land safely, it must be travelling straight down at $4.0\,\text{m s}^{-1}$ or less when it reaches the surface. The maximum acceleration the braking rockets can provide is $1.4\,\text{m s}^{-2}$.

a Determine the minimum length of time the braking rockets would need to be fired to make a safe landing.

_____

_____

_____

b How far from the Moon's surface must the rockets be fired?

_____

_____

_____

5 Two masses, A and B, rest on a frictionless surface, as shown. A force of $40\,\text{N}$ is applied to block A, causing the two masses to accelerate together. Block A has a mass of $6.0\,\text{kg}$ and block B has a mass of $10.0\,\text{kg}$. The questions below require you to use Newton's laws.

a Determine the acceleration of the blocks.

_____

_____

_____

b Determine the force that block A exerts on block B, including the direction of the force.

_____

_____

_____

c What force does block B exert on block A? Justify your response.

_____

_____

_____

6 a Aaron drags a heavy bucket across the ground by applying a $150\,\text{N}$ force to the handle, which is at an angle of $25°$ to the horizontal. Determine the vertical and horizontal components of the force.

_____

_____

b Aaron is now able to apply the same force of 150 N to lift the bucket vertically at a constant velocity to a height of 1.20 m, onto the back of a truck.

    i Calculate the work done in raising the bucket onto the truck.

_____

_____

    ii Determine the gravitational potential energy of the bucket when it is 1.2 m above the ground.

_____

    iii If the bucket was dropped from this height, how much kinetic energy would it have just before it reached the ground?

_____

1 Projectile motion

WS 1.1 Analysing projectile motion by resolving horizontal and vertical components

STUDENT BOOK
Pages 31–54

LEARNING GOALS

Consider the motion of projectiles in terms of two perpendicular components

Analyse projectile motion in terms of uniformly accelerated vertical and constant horizontal motion

A model that resolves the motion of projectiles into independent horizontal and vertical vector components enables projectile motion to be predicted and analysed. The vertical motion is uniformly accelerated, and the horizontal motion is constant.

By resolving vectors for the initial conditions of the projectile and applying equations of motion for the accelerated vertical motion and the constant horizontal motion, predictions can be made about the rest of the motion of the object.

State numerical answers correct to an appropriate number of significant figures.

1 A bullet fired from a gun and another bullet dropped from the same height simultaneously will reach the ground at the same time, although many metres apart. What principle is this illustrating?

_____

_____

2 With the Sun directly overhead, it is observed that the shadow of a ball that is undergoing projectile motion moves with constant speed. What principle of projectile motion is this illustrating?

_____

_____

3 Abiol is at the top of the tall mast of a boat. She drops an apple as the boat sails forward at a constant speed in a straight line on smooth water.

a Where will the apple land on the boat in relation to the mast?

_____

b What path would the apple be observed to take by Philip on the wharf as the boat passes?

_____

_____

4 A stranded sailor is facing north as she fires a flare, at $100\,\mathrm{m\,s^{-1}}$, from a flare gun pointing forwards and up at 75.0° to the horizontal.

a How fast is the flare travelling north?

_____

b How fast is the flare travelling up?

_____

5 a When a projectile reaches its maximum height, what is the value of the y component of its velocity? Explain.

_____

b Use that knowledge to predict when the projectile will have its lowest velocity value during its trajectory. Justify your response.

_____

_____

6 Use the equations for the x and y components of initial velocity and the equation of motion $s = ut + \dfrac{1}{2}at^2$ to show that range depends only on u, θ and t.

_____

_____

7 The motion of a projectile that is launched and lands at the same level is often said to be symmetrical. Use equations of motion to show that, for such a projectile, $v_y = -u_y$ and $t_{up} = t_{down}$. It may help to set out your working in a table.

_____

_____

_____

_____

_____

_____

_____

_____

_____

_____

_____

SB STUDENT BOOK Pages 31–54

Use equations of projectile motion to determine unknown quantities in problems

Derive quantitatively the relationships between various combinations of projectile motion variables, including initial velocity, launch angle, maximum height, time of flight, final velocity, launch height and range, using graphs and equations

When an object is thrown, kicked, fired, hit or launched, the initial conditions of its motion will determine how it moves. Features of its motion, such as range, maximum height, final velocity and time of flight, will be determined by the launch speed and angle, and the launch height compared to the end height.

State numerical answers correct to an appropriate number of significant figures.

1 A ball is launched on flat ground at a speed of $10\,\mathrm{m\,s^{-1}}$ at an angle of $30°$ to the horizontal.

Using appropriate equations, determine:

a the time of flight.

_____

_____

_____

_____

_____

b maximum height.

_____

_____

_____

_____

_____

c the range.

_____

_____

d the final speed of the ball.

_____

9780170449687

2 By calculating the same four quantities as in question **1** for launch speeds of $20\,\text{m}\,\text{s}^{-1}$ and $30\,\text{m}\,\text{s}^{-1}$, determine, in general terms, how launch speed affects each of the four quantities.

_____

_____

_____

_____

_____

_____

_____

_____

3 Consider an object that is launched, and lands, on flat ground. Using a table of values with an arbitrary launch speed of $u = 98\,\text{m}\,\text{s}^{-1}$ and a sketch graph, or by derivation, determine the relationships between launch angle and

a time of flight.

_____

_____

_____

b final vertical velocity.

_____

_____

_____

c range.

_____

_____

_____

9780170449687

d maximum height.

_____

_____

_____

4   The acceleration due to gravity is $9.8\,\text{m s}^{-2}$ on Earth and $1.6\,\text{m s}^{-2}$ on the Moon.

Suppose Alan Shepard had hit identical golf shots with an initial velocity of $50\,\text{m s}^{-1}$ at an angle of $40°$ above horizontal on flat parts of the Moon and Earth. Compare the time of flight, range and maximum height of each shot.

_____

_____

_____

_____

_____

_____

_____

_____

5   **a**   Charlotte wishes to catapult a thumb drive to her colleague Tewolde on the apartment balcony 18 m directly above. What is the minimum launch speed to ensure it reaches Tewolde?

_____

_____

_____

b Having copied the files, Tewolde on the balcony throws the thumb drive at $15\,\mathrm{m\,s^{-1}}$ at $10°$ above the horizontal.

    i How fast will the thumb drive be travelling when it reaches Charlotte on the ground?

    ii How far from the base of the building will Charlotte have to stand to catch the thumb drive?

6 A helicopter is set to release a pallet of fire retardant onto an area affected by bushfire. The helicopter hovers 800 m directly above a point marked X on the field below, as the pallet leaves the helicopter's exit ramp. The ramp slopes at $15°$ below the horizontal and the pallet is travelling at $12\,\mathrm{m\,s^{-1}}$ as it leaves the ramp.

    a How far from the point marked X does the pallet land?

    b How fast will the pallet be moving just before it lands?

7 At the same time as a cricket ball is hit up at $60°$ to the horizontal at $25\,\mathrm{m\,s^{-1}}$, Ellyse sprints straight towards the ball, so that her direction is exactly opposite to the direction of the horizontal velocity component of the ball. If Ellyse was 90 m away from the ball when she started running, at what average speed will she need to run to catch the ball at the same height as that at which it was hit?

2.1 **Understanding the sources of centripetal force (F_c) in various uniform circular motion situations**

STUDENT BOOK
Pages 55–71

LEARNING GOALS

Analyse the forces acting on an object moving in uniform circular motion in a variety of situations with one or two forces acting

Use vector diagrams to analyse uniform circular motion in situations involving two forces

When an object is undergoing uniform circular motion there must be a net force – called the centripetal force – directed towards the centre of the circular path. That centripetal force (F_c) can be the result of a single force acting or a combination of forces.

State numerical answers correct to an appropriate number of significant figures.

1 A pole is secured to the centre of an ice-rink. A rope is tied to the pole with a knot that enables it to move freely around the pole. Tonya is skating in a straight line when she grasps the end of the rope. Holding on to it without any other change, she executes a full circle.

What is the source of the centripetal force causing Tonya to move in a circle in this situation? What would normally have provided the centripetal force enabling her to undergo uniform circular motion if there was no rope?

_____

_____

2 A coin rests on the rubber surface of a record-player turntable, 10 cm from the centre. As long as the turntable's speed is set to 33 revolutions per minute (rpm), the coin will undergo uniform circular motion.

a What is providing the centripetal force to enable the coin to move in a circle in this situation?

_____

b In what way is this situation similar to a car or bicycle moving in a circle around a roundabout?

_____

c In each of the cases above – coin, car or bicycle – what happens if the speed is increased, and why?

_____

_____

_____

_____

3 A hydrogen atom can be modelled as an electron orbiting a proton in uniform circular motion.

What is the source and direction of the centripetal force causing the electron to move in a circle in this situation? What factors determine the magnitude of the centripetal force experienced by the electron?

_____

_____

4 NBN Co-1A is a satellite orbiting approximately 35 786 km above Earth's surface. This satellite has coverage of all of the Australian mainland and many offshore territories.

What is the source and direction of the centripetal force causing the satellite to move in a circle in this situation? What factors determine the magnitude of the centripetal force experienced by the satellite?

_____

_____

5 In the amusement park ride called the rotor, participants are in a circular rotating room that has rough walls and a floor that can be lowered as the rotational speed increases.

Getty Images/Popperfoto

The participants are moving in a circle. What is the source and direction of the centripetal force causing the people to undergo uniform circular motion? Why is it possible to lower the floor and leave participants 'stuck' to the wall, as shown, only when the rotational speed is sufficiently high?

_____

_____

_____

_____

6 A rollercoaster is a common feature of amusement parks. The motion of the rollercoaster in a vertical circle can be modelled as uniform circular motion.

Shutterstock.com/Martin Charles Hatch

9780170449687

Using labelled diagrams, compare the forces on the rollercoaster at the top of the loop with those at the bottom.

_____

_____

_____

_____

7 A mass (m) can be made to move in a horizontal circle of radius r on the end of a string of length L, as seen in the diagram below. This set-up is called a conical pendulum.

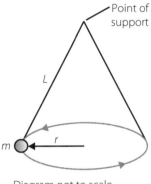

Diagram not to scale

On the diagram, label the two forces acting on the mass. Construct a vector diagram to show how these forces add to create the centripetal force that causes the mass to move in a circle.

8 Vehicles can navigate a banked circular corner safely at speeds that are higher than the speeds that are safe on a flat corner of the same radius. In addition to the frictional force, the bank provides centripetal force as a consequence of two forces acting on the car.

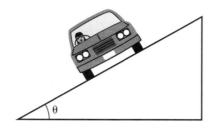

Label these two forces on the diagram. Construct a vector diagram to show how these forces add to create the centripetal force that helps the car to move in a circle in this situation.

WS **2.2** **Problem solving in uniform circular motion using key equations**

Determine unknown quantities related to objects in uniform circular motion

Solve problems by applying mathematical relationships:

$$a_c = \frac{v^2}{r}, v = \frac{2\pi r}{T}, F_c = \frac{mv^2}{r}, \omega = \frac{\Delta \theta}{t}$$

HINT

These questions build on the situations described in Worksheet 2.1. Looking back to those questions and answers may assist with these questions.

State numerical answers correct to an appropriate number of significant figures.

1   A pole is secured to the centre of an ice-rink. A rope is tied to the pole with a knot that enables it to move freely around the pole. Tonya is skating in a straight line at $12.0\,\mathrm{m\,s^{-1}}$ when she grasps the end of the rope. Holding on to it without any other change, she executes a full circle of radius $10.0\,\mathrm{m}$.

   Calculate the centripetal acceleration (a_c) for Tonya moving in a circle in this situation.

   _____

   _____

2   A coin rests on the rubber surface of a record-player turntable, $10\,\mathrm{cm}$ from the centre. As long as the turntable's speed is set to 33 revolutions per minute (rpm), the coin will undergo uniform circular motion.

   By first calculating the time taken for the coin to move through a circle, determine its speed (v).

   _____

   _____

3   A hydrogen atom can be modelled as an electron orbiting a proton in uniform circular motion. The angular velocity (ω) of this motion is published as $4.11 \times 10^{16}\,\mathrm{radian\,s^{-1}}$.

   Remembering that one revolution is 2π radians, how long will it take the electron to complete one billion (10^9) orbits of the proton?

   _____

   _____

4   The Westpac weather satellite orbits Earth at a height of $100\,\mathrm{km}$ above Earth's surface. Earth has a radius of $6370\,\mathrm{km}$ and the satellite has a speed of $7850\,\mathrm{m\,s^{-1}}$.

   a   Calculate the centripetal acceleration.

   _____

   _____

b Determine the time taken to complete one orbit. Express your answer in minutes and seconds.

_____

_____

c Calculate the angular velocity and determine the relationship between speed and angular velocity.

_____

_____

5 In the amusement park ride called the rotor, participants rely on friction to provide an upwards force to balance the gravitational force (mg).

Recall from Year 11 that the force of friction is given by the equation $F_{friction} = \mu F_N$.

Using the information above, and your understanding of the source of the centripetal force in the rotor, generate an equation linking weight force to centripetal force. Use this equation to determine the minimum coefficient of friction (μ) necessary to ensure the rotor works effectively when the velocity of participants is $8.5\,\mathrm{m\,s^{-1}}$ and the radius is $5\,\mathrm{m}$.

_____

_____

_____

6 The motion of a rollercoaster in an amusement park can be modelled as uniform circular motion. In this situation, all the centripetal force at the top of the circle is provided by gravity. Determine the minimum speed required for a rollercoaster to complete a circle of radius $30\,\mathrm{m}$ in uniform circular motion.

_____

_____

_____

7 A conical pendulum, such as the one illustrated below, is set in motion on a string of length $1.05\,\mathrm{m}$ and with an orbital radius of $10\,\mathrm{cm}$.

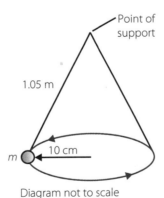

Point of support

1.05 m

10 cm

m

Diagram not to scale

a Determine the angle that the string makes with the vertical.

_____

b Using a vector diagram of the forces acting on the mass, show that $\tan\theta = \dfrac{v^2}{rg}$ and determine the speed at which the mass moves.

_____

_____

_____

_____

c Calculate the period of the pendulum.

_____

_____

8 A circular corner is banked at an angle that enables it to be navigated safely by vehicles at $108\,\mathrm{km\,h^{-1}}$ without relying on any friction. The radius of the corner is $1100\,\mathrm{m}$.

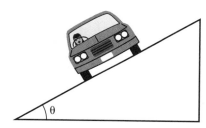

Using a vector diagram of the forces acting on the car, show that $\tan\theta = \dfrac{v^2}{rg}$ and determine the angle at which the bank must be inclined.

_____

_____

_____

Understanding factors affecting torque through working scientifically skills

STUDENT BOOK
Pages 74–76

LEARNING GOALS

Investigate the relationship in a rotating system between the applied force, the angle at which the force is applied, the distance to the pivot and the torque that results

The torque, τ, acting on any rotating system is a function of three variables: the applied force (F), the angle at which that force is applied (θ), and the distance (r) from the pivot at which that force is applied. This is summarised in the equation $\tau = rF\sin\theta$.

State numerical answers correct to an appropriate number of significant figures.

1 Lai gathers data during an experiment to investigate relationships between torque and force. She keeps distance and angle constant and uses a precise device to measure the torque applied in an experimental set-up similar to the one in the diagram below.

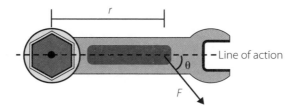

She plots her collected data and draws a line of best fit, as shown.

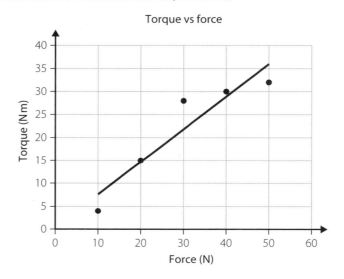

a What does the graph indicate about the relationship between τ and F?

_____

b Using the graph, comment on the reliability of the collected data. Justify your comment.

_____

_____

c Using the line of best fit, comment on the accuracy of the collected data. Justify your comment.

_____

_____

d Using the gradient of the line of best fit, determine the product of distance and the sine of the angle ($r\sin\theta$).

_____

_____

2 Lai uses the same set-up to investigate the relationship between torque (τ) and distance (r). This time she keeps F and θ constant. Again, she plots her data and draws a line of best fit.

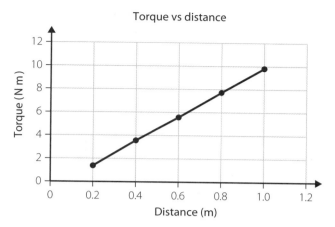

Torque vs distance

a What does the graph indicate about the relationship between τ and r?

_____

b Using the graph, comment on the reliability of the collected data. Justify your comment.

_____

_____

c Using the line of best, comment on the accuracy of the collected data. Justify your comment.

_____

_____

d Using the gradient of the line of best fit, determine the product of force and the sine of the angle ($F\sin\theta$).

_____

_____

3   To investigate torque vs angle (τ vs θ), Lai keeps force and distance constant. She records her data in the following table.

Angle θ (°)	Torque (N m)	
15	1.3	
30	2.5	
45	3.5	
60	4.3	
75	4.8	
90	5	

a   Plot the data on the axes provided below, with angle on the horizontal axis and torque on the vertical axis. Sketch a *curve* of best fit for the plotted data.

Explain how the shape of your graph confirms the relationship between angle and torque that is suggested by the torque equation.

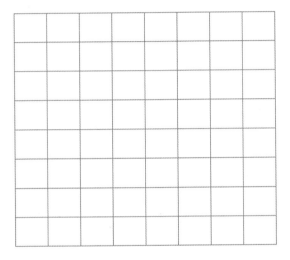

b   Complete the blank column of the table with values of sin θ. Use 3 significant figures for precision.

c   By constructing an appropriate graph on the axes below, use the line of best fit to graphically determine the value of force times distance (Fr).

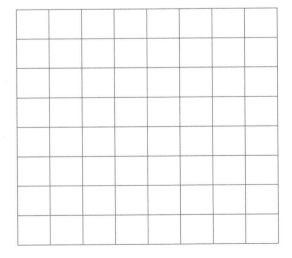

3 Motion in gravitational fields

 3.1 Understanding the relationships between variables in the motion of planets and satellites

STUDENT BOOK
Pages 84–91

LEARNING GOALS

Analyse quantitative aspects of the orbital motion of planets and satellites

Use the mathematical relationships between gravitational force, centripetal force, centripetal acceleration, mass, orbital radius, orbital velocity and orbital period to determine unknown quantities

Planets and satellites can be considered to be undergoing uniform circular motion. The net force acting on the object – called the centripetal force – is directed towards the centre of the circular path and is provided by the gravitational force.

HINT

Note that $km\,h^{-1}$ must be converted to $m\,s^{-1}$ (by dividing by 3.6), km must be converted to m (by multiplying by 1000) and time values must be converted to seconds throughout this section.

State numerical answers correct to an appropriate number of significant figures.

1 Determine the centripetal acceleration of the Polar V satellite, which moves at a speed of $26\,528\,km\,h^{-1}$ at a distance of 1000 km above Earth's surface. Radius of Earth is 6370 km.

_____

_____

2 Compare the centripetal force exerted by the Sun on Mercury (mass $3.30 \times 10^{23}\,kg$, orbital speed $47.4\,km\,s^{-1}$ and orbital radius 57.9 million km) with the centripetal force exerted by the Sun on Pluto (mass $1.46 \times 10^{22}\,kg$, orbital speed $4.7\,km\,s^{-1}$ and orbital radius 5910 million km).

_____

_____

3 Phobos, one of the moons of Mars, has an orbital radius of 9377 km and orbital period of 7.66 hours. Determine its orbital velocity.

_____

_____

4 Determine the orbital period of Earth in days, given the published data for Earth is that it moves at $107\,280\,km\,h^{-1}$ and is an average of 149.6 million km from the Sun.

_____

_____

5 Calculate the force due to gravity acting on a physicist of mass 75 kg on a research station in orbit 1800 km above Earth. Mass of Earth is 6.0×10^{24} kg.

_____

_____

6 By equating centripetal force and force due to gravity for a mass orbiting Earth, show that orbital velocity and

orbital radius are related by the equation $v = \sqrt{\dfrac{GM_{\text{Earth}}}{r}}$.

_____

_____

7 a Complete the table below regarding satellites orbiting Earth. (Note that data values have not been rounded consistently.)

Satellite	m (kg)	r (m)	T (s)	v (m s^{-1})	a_c (m s^{-2})	F_c (N)	F_g (N)
ISS	420 000			7.66×10^3			2.46×10^{13}
Moon	7.35×10^{22}	3.78×10^8					
Optus	1350	7.85×10^6		7140			
Optus D3		6.87×10^6			8.48		20 521
Oscar	500	7.37×10^6	6284				
Sky Muster	875		5992			6869	
Wresat		4.2164×10^7		3081		211.5	

b Explain why the values in the last two columns of the table above must be the same for all satellites.

_____

_____

_____

Exploring aspects of satellite motion and how these relate to use

STUDENT BOOK
Pages 86–90

Make predictions about the orbital properties of satellites in a variety of situations

Analyse properties of satellites near the Earth and in geostationary orbits, and relate these to their uses

Artificial satellites are placed into orbit to fulfil a variety of roles. The orbital radius and period of the satellite dictate how it can be used. These variables are both linked to the speed of the satellite. Most satellites of interest are either low-Earth orbit satellites (LEOs) or geostationary satellites (GEOs).

State numerical answers correct to an appropriate number of significant figures.

1 LEOs orbit relatively close to the surface of the Earth, at altitudes typically between about 160 km and 1000 km. Outline three uses for satellites orbiting this close to Earth and explain why this short distance is helpful.

_____

_____

_____

2 The Hubble Space Telescope is a LEO that orbits 540 km above Earth. Use the equation $v = \sqrt{\dfrac{GM_{Earth}}{r}}$ to determine the Hubble Space Telescope's orbital velocity.

_____

_____

3 Use the result from Question 2 to calculate the orbital period of the Hubble Space Telescope.

_____

_____

4 As their category suggests, all geosynchronous satellites have an orbital period exactly equivalent to the period of rotation of Earth on its axis – 23 hours, 56 minutes and 4 seconds (86 164 s). The equation below relates orbital period (T) to orbital radius (r). Use this equation to determine the altitude of all geosynchronous satellites. (This equation will be explored in more detail in subsequent worksheets.)

$$\frac{r^3}{T^2} = \frac{GM}{4\pi^2}$$

_____

_____

_____

5 Compare the orbital velocity of a geosynchronous satellite to that of the Hubble Space Telescope (from Question 2).

_____

_____

6 List three uses of geosynchronous satellites and relate these uses to their distance from Earth's surface.

_____

_____

_____

7 Geostationary satellites are special geosynchronous satellites that orbit directly above the equator. If these satellites were able to be observed from Earth, describe how an observer on Earth would see a geosynchronous satellite move compared to a geostationary satellite.

_____

_____

_____

_____

_____

8 Many LEOs can be seen by observers on Earth because they reflect sunlight and are relatively close. Describe how the ISS might appear to an observer in eastern Australia, for example, over a series of nights.

_____

_____

_____

_____

_____

Understanding how Kepler's laws relate to the motion of planets and satellites

Understand Kepler's Laws of Planetary Motion

Understand the total energy of planets in orbit

Apply the following relationships to planetary motion:

$$v = \frac{2\pi r}{T} \qquad \frac{r^3}{T^2} = \frac{GM}{4\pi^2}$$

Johannes Kepler provided several key laws that aid our understanding of planets and satellites undergoing uniform circular motion, or in elliptical orbits, under the influence of a net force provided by gravity.

State numerical answers correct to an appropriate number of significant figures.

1 State Kepler's first law.

_____

2 With the aid of a labelled diagram, outline the key geometrical aspects of an ellipse.

_____

_____

3 Eccentricity is a number that indicates the degree to which an ellipse is elongated. Numerically it is the ratio of the distance from the centre to the focus to the distance from the centre to the vertex. The eccentricity of a circle (a special kind of ellipse), therefore, is 0. Values of the eccentricity of some planets of our solar system are listed in the table below. Discuss these values with particular reference to the orbits of Earth and Mercury.

Planet	Eccentricity
Venus	0.0068
Earth	0.0167
Jupiter	0.0484
Saturn	0.0541
Mars	0.0934
Mercury	0.2056

_____

_____

_____

_____

_____

4 Kepler's second law is known as the Law of Areas. It states that each planet 'sweeps out' an equal area during equivalent time intervals of its orbit. Assuming that the planet in orbit is a closed system (no energy is added or removed), then conservation of energy applies to the planet as it moves through its orbit. Contrast the kinetic energy and gravitational potential energy of the planet at the closest point to the point to where it is furthest from the Sun, and use this to justify Kepler's second law.

_____

_____

_____

_____

_____

5 Kepler's third law is given as

$$\frac{r^3}{T^2} = \frac{GM}{4\pi^2}$$

where M is the mass of the orbited body.

What does this imply for the ratio $\dfrac{r^3}{T^2}$ for all planets in the solar system? Justify your answer.

_____

_____

6 A team of astronomers gathered data about the orbital radius, in metres, and orbital period, in seconds, of four planets of our solar system. The data was then manipulated to be plotted on the graph below.

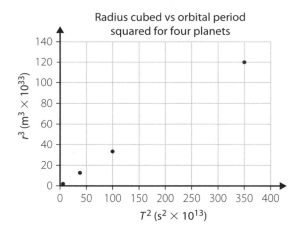

a Use the graph to determine the mass of the Sun in kg.

_____

_____

_____

b Comment on the accuracy of the value determined graphically given the published value of 1.989×10^{30} kg.

_____

_____

_____

 9780170449687

Using gravitational force and gravitational potential energy to understand more about the motion of planets and satellites

STUDENT BOOK
Pages 94–98

LEARNING GOALS

Derive equations seen below for escape velocity and total energy (U + K)

Calculate values of escape velocity, potential energy, kinetic energy and total energy for a variety of situations

Apply Kepler's Third Law of planetary motion:

$$v_{esc} = \sqrt{\frac{GM}{r}} \qquad U = -\frac{GMm}{r} \qquad U + K = -\frac{GMm}{2r}$$

HINT

Note that only one of the equations above appears in the Data and Formulae Sheet. The others may be expected to be derived.

Planets and satellites move in radial gravitational fields. Consequently, they have both kinetic energy, as a consequence of their motion, and gravitational potential energy, as a consequence of their position in the field.

State numerical answers correct to an appropriate number of significant figures.

1 Complete the table below for planets orbiting the Sun ($M = 1.989 \times 10^{30}$ kg). Note that data values have not been rounded consistently.

Planet	r (m) $\times 10^9$	T (days)	m (kg) $\times 10^{24}$	U (J) $\times 10^{32}$	E_{Total} (J) $\times 10^{32}$	v_{escape} (km s^{-1})
Mercury	57.9			−7.553		
Venus	108.2				−29.93	
Earth						11.2
Mars				−3.748		5.0
Jupiter					−1625	
Saturn		10 747				
Uranus		30 581		−40.18		
Neptune					−15.15	23.5

HINT

Values in the table may vary slightly depending on the method of calculation. Significant figures have not been applied, in order to provide more precise data.

2 Planets and satellites move in uniform circular motion under the influence of a centripetal force provided by the gravitational force between the orbiting object (mass m) and the orbited object (mass M). By equating F_c and F_g and substituting in the equation relating v, r and T from preceding worksheets, derive Kepler's third law, the Law of Periods:

$$\frac{r^3}{T^2} = \frac{GM}{4\pi^2}$$

3 a Show, with reference to the Universal Law of Gravitation, why it is assumed that the gravitational force exerted by one object (M) on another object (m) is zero when the objects are separated by an infinite distance ($r = \infty$).

b Given that gravitational potential energy (U) is the energy of an object experiencing a force in a gravitational field, what then must be the value of U at this infinite distance? Justify your response.

c It is then assumed that at all positions closer than an infinite distance the value of U must be negative. How can this assumption be justified?

4 In Question 6 of Worksheet 3.1, you equated centripetal force and force due to gravity for a mass m orbiting another mass M to show that orbital velocity and orbital radius are related by the equation $v = \sqrt{\dfrac{GM}{r}}$.

a By substituting this expression for v into the equation for kinetic energy, derive an expression for kinetic energy of an orbiting object in terms of G, M, m and r.

b Show that the total energy of an orbiting planet or satellite can be given by the equation $E_{\text{total}} = -\dfrac{GMm}{2r}$.

5 On the axes below, sketch labelled curves for each of U, K and E_{total} for a satellite at various orbital radii.

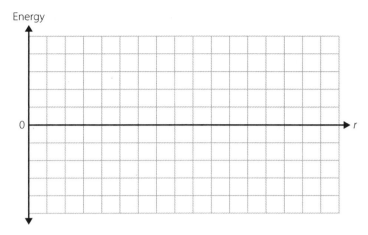

6 Calculate the minimum amount of energy expended by the thruster of the International Space Station ($m = 420\,000$ kg) to move it from an orbit of altitude 340 km to an orbit of altitude 420 km.

_____

_____

_____

_____

7 Escape velocity is a hypothetical initial speed at which an object must be projected away from a planet to permanently escape that planet's gravitational field. Permanent escape quantitatively means reaching an infinite distance as it comes to rest.

a Show that the kinetic and gravitational potential energy values at that point would both be zero.

_____

_____

_____

b Conservation of energy requires that the total energy of the object when it was projected from Earth's surface must also have been zero. Use this fact to derive the expression for escape velocity.

_____

_____

_____

_____

Module five: Checking understanding

Circle the correct answer for questions **1–3**.

1 A car travels around two separate bends that have the same radius. Bend A is a flat bend. Bend B is a banked curve. The car takes both bends at high speed.

Which of the following is correct about the forces involved?

A All of the centripetal force on Bend B is provided by the sideways frictional force of the tyres.

B Part of the centripetal force on Bend A is provided by the sideways frictional force of the tyres.

C All of the centripetal force on Bend B is provided by the horizontal component of the normal force.

D The centripetal force provided by the sideways frictional force from the tyres must be greater on Bend A than on Bend B.

2 Two satellites, A and B, are orbiting Earth at different altitudes, as shown.

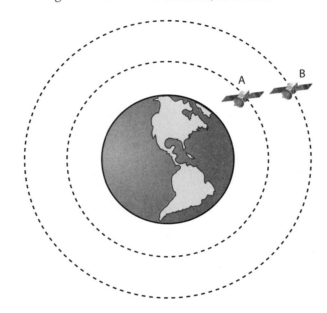

Which comparison of the satellites' orbital velocities, orbital periods and centripetal accelerations is correct?

	Orbital velocity	Orbital period	Centripetal acceleration
A	A is greater	B is greater	B is greater
B	B is greater	A is greater	A is greater
C	B is greater	B is greater	B is greater
D	A is greater	B is greater	A is greater

3 The ratio $\dfrac{r^3}{T^2}$ is the same for:

A all planets in the solar system.

B only the inner planets of the solar system.

C all moons of all planets in the solar system.

D all planets and all moons in the solar system.

State numerical answers correct to an appropriate number of significant figures.

4   A golfer strikes the ball at 30° to the horizontal and hits a 'hole-in-one' 135 m away, as shown in the diagram below.

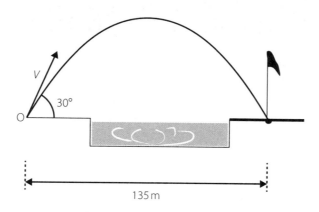

Calculate the magnitude of the initial velocity of the ball.

_____

_____

_____

5   A car is travelling around a frictionless banked track, angled at 15.0°, and having a turning radius of 26.0 m, as shown in the diagram. The centripetal force acting on the car is 30 400 N. Determine the mass of the car.

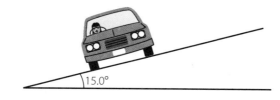

_____

_____

_____

6   A satellite orbits a recently discovered planet. The satellite has an orbital velocity of 8.167 m s^{-1} and a period of 3486.7 s. Determine:

a   the gravitational field strength in the region of the satellite's orbit.

_____

_____

_____

b   the mass of the planet.

_____

_____

_____

_____

Reviewing prior knowledge

State numerical answers correct to an appropriate number of significant figures.

1   A pair of parallel plates is shown. One is positively charged and the other is negatively charged.

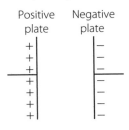

Positive plate   Negative plate

a   On the diagram, sketch an appropriate set of electric field lines.

b   List three or more important characteristics of electric field lines.

_____

_____

_____

_____

_____

c   Assume the plates are 2.5 cm apart and the potential difference between them is 175 V. Calculate the magnitude of the electric field between the plates and indicate its direction.

_____

_____

_____

d   Calculate the kinetic energy gained by an electron when it accelerates from the negative to the positive plate. Write your answer in electron volts and in joules.

_____

_____

_____

_____

_____

e Assuming it started from rest, calculate the speed of the electron just before it hits the positive plate.

_____

_____

_____

_____

f What is the equation for the force on the electron of the electric field? Write it using vector notation.

_____

g Because the field is constant, the force, and therefore the acceleration, will be constant. Use the constant acceleration equations to work out how long the electron takes to cross the gap between plates.

_____

_____

_____

_____

_____

_____

2 a State Ohm's Law and define the variables.

_____

_____

b A voltage of 232 V is applied across a length of electrical heating wire with resistance 25 Ω. Determine the current in the wire.

_____

_____

_____

c How much power is being consumed by the heating wire?

_____

_____

3 Describe, with a supporting diagram, the use of the right-hand rule to determine the direction of the magnetic field due to:

a a long straight wire.

_____

_____

_____

b a loop of wire.

_____

_____

_____

WS 4.1 Exploring the forces charged particles experience in electric fields and how this changes their energy

STUDENT BOOK
Pages 106–111

LEARNING GOALS

Explore, quantitatively, interactions between charged particles and uniform electric fields

Fields are all around us. Our control and use of electric and magnetic fields are important aspects of modern technology. A charged particle in an electric field behaves a lot like a mass in a gravitational field. Magnetic fields have a different effect.

State numerical answers correct to an appropriate number of significant figures.

1   A proton (p), a neutron (n) and an electron (e) sit side by side in a uniform electric field of $100\,\mathrm{V\,m^{-1}}$ downwards, as shown. (Ignore the forces that they might exert on each other.)

Deduce the direction of the force due to the electric field felt by:

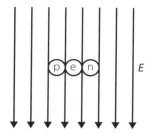

a   the proton.

_____

b   the electron.

_____

c   the neutron.

_____

2   Calculate the magnitude of the force, due to the electric field, experienced by:

a   the proton.

_____

_____

_____

b the electron.

_____

_____

_____

c the neutron.

_____

3 Calculate the magnitude of the acceleration, due to the electric field, of:

a the proton.

_____

_____

_____

_____

b the electron.

_____

_____

_____

_____

c the neutron.

_____

4 Each particle starts at rest. Using $W = \Delta K = qEd$, calculate the final kinetic energy, after moving 1.00 mm, of:

a the proton.

_____

_____

_____

_____

b the electron.

_____

_____

5 Explain why the calculation in Question **4** was not done for the neutron.

_____

_____

6 The uniform electric field strength is $100\,V\,m^{-1}$ downwards. If the electrical potential is $250\,V$ at some height, determine the electrical potential:

   a at a point $2.5\,m$ higher.

_____

_____

_____

_____

   b at a point $2.5\,m$ lower.

_____

   c at a point $2.5\,m$ to the right.

_____

   d at a point $2.5\,m$ to the right and $3.5\,m$ higher.

_____

7 Using concepts of work, kinetic and potential energy, how does the energy of the proton from the questions above change as it moves? Explain why.

_____

_____

_____

8 Using work, kinetic and potential energy, how does the energy of the electron from the questions above change as it moves? Explain why.

_____

_____

_____

Understanding how charged particles move in electric fields

Quantitatively analyse the motion of charged particles in electric fields

Newtonian mechanics – such as $\vec{F}_{net} = m\vec{a}$ – applies just as much for electric fields as for gravitational fields. That means we can use our existing knowledge of vectors and the constant acceleration equations to predict how charged particles will behave in an electric field.

State numerical answers correct to an appropriate number of significant figures.

1 A proton, a neutron and an electron each have an initial horizontal velocity to the right of $112\,\text{m s}^{-1}$, in a uniform electric field of $100\,\text{V m}^{-1}$ downwards (see figure). Ignore gravity.

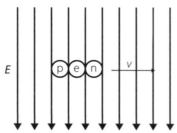

    a Do any of the particles experience a horizontal acceleration? Justify your answer. If your answer is 'yes' in any case, indicate the direction of the acceleration.

    b Do any of the particles experience a vertical acceleration? Justify your answer. If your answer is 'yes', indicate the direction of the acceleration.

2 Calculate the vertical and horizontal velocity components, after freely accelerating for 1.00 ns, of:

    a the proton.

    b the electron.

    c the neutron.

3 Using the results from Question **2**, calculate the direction and magnitude of the velocity of each particle after 1.00 ns. Consider using a vector diagram.

a _____

_____

_____

_____

_____

_____

_____

_____

_____

_____

b _____

_____

_____

_____

_____

_____

c _____

_____

_____

4 Sketch a *qualitative* graph of *vertical* position versus time for the proton for the first 2.00 ns.

Analysing how charged particles move in magnetic fields

Analyse the interaction between charged particles and uniform magnetic fields and solve problems

Knowledge of mechanics can be used to model how charged particles behave in magnetic fields. The magnetic force acts perpendicularly to the particle velocity, often resulting in circular motion.

State numerical answers correct to an appropriate number of significant figures.

1 A proton with velocity $v = 1100\,\mathrm{m\,s^{-1}}$ is moving through a uniform magnetic field of strength 4.00 T.

   a Calculate the magnitude of the force experienced by the proton if the magnetic field is parallel to the velocity.

   _____

   b Calculate the magnitude of the force experienced by the proton if the magnetic field is perpendicular to the velocity.

   _____

   _____

   _____

2 Assuming they are all travelling at the same speed perpendicular to the same magnetic field, which particle would experience the greatest acceleration – a proton, a neutron, an alpha particle or an electron? Explain your answer. Use the forms of the key equations to assist you, but do not do any calculations.

   _____

   _____

   _____

   _____

   _____

3 Imagine we have a lot of particles of different sizes and charges moving at the same speed in a stream perpendicular to the same uniform magnetic field.

   a Use $a_c = \dfrac{v^2}{r}$ and $F = qvB\sin\theta$ to show how the radius of the curved path of the beam of particles depends on the charge-to-mass ratio, $\dfrac{q}{m}$.

   _____

   _____

   _____

   _____

   _____

   _____

   _____

   _____

b Explain how this might be used to separate particles of different charge-to-mass ratio. (This is the basis of the mass spectrometer, a widely used scientific instrument. For example, mass spectrometers are used at airports to detect traces of explosives on luggage and clothing.) A sketch may be useful.

_____

_____

_____

_____

_____

c Would such a machine be able to easily separate a deuteron (a nucleus of one proton plus one neutron) from an alpha particle? (An alpha particle is a radioactive decay particle consisting of two protons and two neutrons.)

_____

_____

_____

_____

_____

4 In two or three sentences, explain why a magnetic field does no work on a freely moving charged particle. You may find it useful to refer to an equation, but do no calculations.

_____

_____

_____

Extension

5 A charged particle is moving in a uniform magnetic field. The particle has an initial velocity that is neither parallel nor perpendicular to the field lines. Describe the resulting motion, and explain how it arises.

_____

_____

_____

_____

_____

LEARNING GOALS

Analyse the motion of charged objects in electric fields, magnetic fields and situations that combine both types of field

A combination of magnetic and electric fields is often used to control charged particles, and nearly everything is done within a gravitational field. The biggest machine on Earth – the Large Hadron Collider – is basically a device to use these fields to control charged particles – and make them collide.

State numerical answers correct to an appropriate number of significant figures.

1 A drop of oil of mass 0.23 mg and charge −11.3 nC is in an evacuated chamber close to the surface of Earth.

   a Calculate and describe the magnitude and direction of an electric field that would balance the force of gravity on the droplet such that it does not accelerate up or down.

   b Calculate and describe the magnitude and direction of a magnetic field perpendicular to the velocity that would balance the force of gravity on the droplet if it is travelling at $1000\,\mathrm{m\,s^{-1}}$ horizontally.

2 In a static electricity experiment, a ping-pong ball of mass 2.7 g is given a positive charge using a Teflon rod. Astrid wants it to hover against gravity, so she puts it between two large, flat horizontal metal plates. She finds that the ball hovers when the plates are 50 cm apart and the potential difference between them is 20 kV. Deduce the charge on the ball.

9780170449687

3 Below is a diagram of a device called a cyclotron, which is designed to bring charged particles to high speeds. As shown, it uses a combination of magnetic and electric fields in a vacuum chamber to achieve this goal as the particle moves along the illustrated path.

Noting that inside a hollow conductor there is no electric field if no charge is enclosed, explain the role of the electric and magnetic fields in achieving the desired path and speed increase for the charged particle.

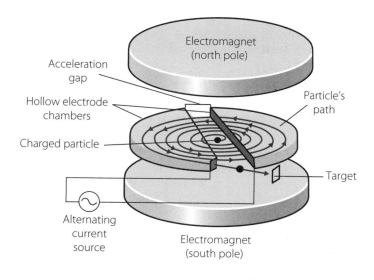

_____

_____

_____

_____

_____

_____

4 Use equations to describe a situation in which it is possible to make a moving charged particle travel in a straight line using a combination of electric and magnetic fields. What would be the relationship between the speed of the charged particles, the electric field strength and the magnetic field strength?

_____

_____

_____

_____

INQUIRY QUESTION: UNDER WHAT CIRCUMSTANCES IS A FORCE PRODUCED ON A CURRENT-CARRYING CONDUCTOR IN A MAGNETIC FIELD?

WS **5.1** ## Exploring how current-carrying wires interact with magnetic fields

STUDENT BOOK
Pages 132–139

LEARNING GOALS

Explore, quantitatively and qualitatively, the behaviour of a current-carrying conductor in a magnetic field using $F = lIB \sin \theta$ and determine how these factors affect the force

A moving charge in a magnetic field experiences a force. A current in a conductor consists of moving charges, so a current-carrying conductor in a magnetic field experiences a force. This is the basis of the electric motor. Because this allows understanding of a relationship between current and mechanical force, it leads to a definition for the ampere, the unit of current.

State numerical answers correct to an appropriate number of significant figures.

1 A current-carrying wire is lying parallel with a magnetic field. Sketch a qualitative graph of the force on the wire as a function of angle as the wire is rotated until it is perpendicular to the field.

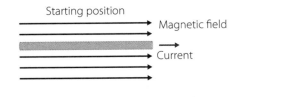

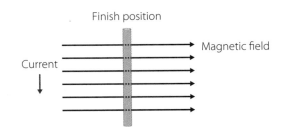

2   Theo positions a horseshoe magnet such that the poles point upwards and the field lines go left to right horizontally. There is a small region between the north and south poles where the magnetic field is uniform and of magnitude 0.0550 T. Theo then dangles a 1.00 cm length of wire in the field, as shown.

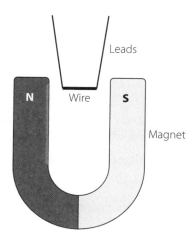

a   The wire is parallel to the magnetic field lines – it points from N to S. There is no current in the wire and the wire is made of copper, which is not ferromagnetic. Deduce the force on the wire due to the magnetic field.

_____

b   The wire remains where it is, but the leads are connected to a power supply and a current on 250 mA flows through the wire. Deduce the force on the wire due to the magnetic field.

_____

c   Theo disconnects the power supply then rotates the wire so that it is perpendicular to the magnetic field lines. (Imagine the same figure but with the wire pointing out of the page.) If the current in the wire is still 250 mA, now coming out of the page, calculate the force on the wire and deduce its direction.

_____

_____

_____

_____

d   Theo disconnects the power supply then rotates the wire so that it is perpendicular to the magnetic field lines, but now it is running up and down the page, with the current flowing from top to bottom. He reconnects the power supply and 250 mA current. Calculate the force on the wire and deduce its direction.

_____

_____

Analysing the forces current-carrying wires exert on each other

STUDENT BOOK
Pages 139–145

Because a current in a wire both creates and responds to a magnetic field, current-carrying wires brought together experience a mutual force that depends on the separation, the currents and the mutual orientations of the wires. If we have two wires that are parallel and both of length l, we can say that $\dfrac{F}{l} = \dfrac{\mu_0}{2\pi} \dfrac{I_1 I_2}{r}$ where r is the distance between the wires. This leads to a definition of the ampere, the unit of electric current.

State numerical answers correct to an appropriate number of significant figures.

1   Sketch a graph to illustrate how the perpendicular distance from the centre of the wire affects the magnitude of the magnetic field strength due to a current-carrying wire. Take the wire radius as R and plot for $r > R$.

2   Explain why 'force per unit length' is a useful quantity.

_____

_____

_____

3   Two long, parallel, vertical wires carrying currents of 1.0 A up and 2.5 A down respectively are 0.50 m apart. Calculate the magnitude of the force per unit length experienced by the wires. Is it attractive or repulsive?

_____

_____

_____

_____

_____

4 Is the rule $F_{\text{by 2 on 1}} = -F_{\text{by 1 on 2}}$ still true if one of the wires is fixed and heavy and cannot move and the other wire is light and free to respond to the force?

_____

5 Explain why the definition of the ampere uses force per unit length rather than simply force.

_____

_____

_____

6 Electromagnetic induction

WS 6.1 Describing magnetic flux and how it can change

STUDENT BOOK
Pages 150–153

LEARNING GOALS

Describe magnetic flux in relation to magnetic field strength and area

Understand what it means for the flux through an area to change with time

Magnetic flux is a measure of how much magnetic field is passing through a given area. Its symbol is capital phi, Φ. To understand important applications of magnetic flux, such as its application in transformers and generators, magnetic flux must be clearly understood. Flux, magnetic field and area are connected by the equation $\Phi = BA\cos\theta$, where A is the area and θ is the angle between the field, B, and the normal to the area.

State numerical answers correct to an appropriate number of significant figures.

1   A loop of wire is suspended in a uniform magnetic field. Using the equation $\Phi = BA\cos\theta$, explain three different ways the flux through the loop can be changed.

_____

_____

_____

2   A square loop of wire of fixed size is rotating in a uniform, constant magnetic field of 1.0 T, as shown below.

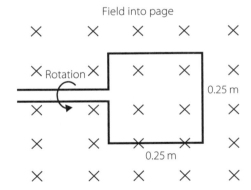

a   Sketch a graph of the flux through the loop as a function of the angle of rotation, where the angle is zero when the field lines are parallel to the normal. Sketch for an angular range $-90° \le \theta \le 180°$.

b The loop is a square of side length 0.25 m. Calculate the magnetic flux when the field is parallel to the normal to the area.

_____

_____

_____

c Calculate the flux after the loop is rotated such that field is at an angle of 34° to the normal to the area.

_____

_____

_____

d Calculate the flux after the loop is rotated such that field is at an angle of 90° to the normal to the area.

_____

_____

3 The diagram below shows a rectangular current loop perpendicular to a uniform magnetic field into the page. The loop is completed by the sliding conducting metal bar on the right. An ammeter is placed in the circuit.

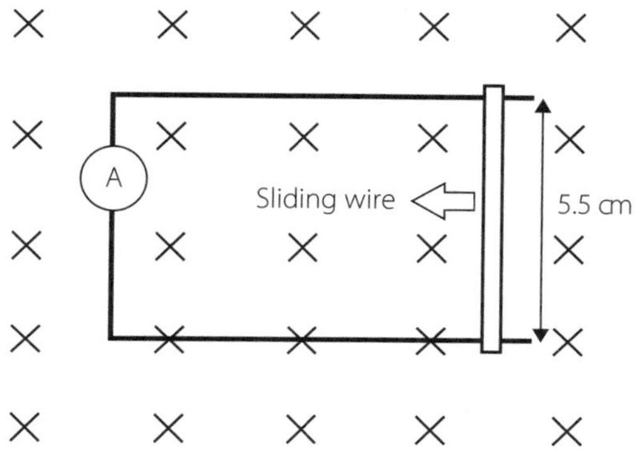

If the magnetic field is of magnitude 0.25 T and the bar moves at 0.50 m s^{-1}, what is the rate of change of the flux?

_____

_____

_____

_____

_____

_____

_____

_____

_____

_____

_____

_____

Analysing electromagnetic induction

Understand the process of electromagnetic induction through Faraday's Law and Lenz's Law

When the flux through a loop of wire changes, a current flows through the loop. This implies that an emf, ε, is generated. The faster the flux changes, the bigger the emf. The relationship is summed up by Faraday's Law:

$$\varepsilon = -N\frac{\Delta\Phi}{\Delta t} = \frac{-\Delta(NBA\cos\theta)}{\Delta t}$$

where Δ means 'change in'; for example $\Delta\Phi = \Phi_{final} - \Phi_{initial}$.

The change in flux induces an emf that causes a current to flow in the wire. In turn, the current in the wire generates its own magnetic field. The negative sign in the equation says that the current created in the wire loop acts to create a magnetic field that opposes the change in flux. For example, if the flux is decreasing, the induced current will create a field to increase the flux. This is Lenz's Law.

State numerical answers correct to an appropriate number of significant figures.

1   The graph below shows the magnetic flux (in units of weber, Wb) passing through a loop of wire, plotted as a function of time.

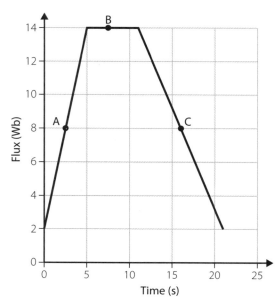

a   Determine the magnitude and sign of the induced emf at time A.

_____

_____

b   Determine the magnitude and sign of the induced emf at time B.

_____

_____

c   Determine the magnitude and sign of the induced emf at time C.

_____

_____

d Sketch a graph of the induced emf as a function of time.

e Explain how your results to **a**, **b**, **c** and **d** would change if, instead of the loop, a coil of 15 turns of wire is used.

_____

_____

_____

2 In Question 3 of Worksheet 6.1, you determined the rate of change of the flux in a loop of wire.

   a Use Faraday's Law to calculate the resulting emf.

_____

_____

_____

_____

   b Use Lenz's Law to deduce the direction of the current – clockwise or counterclockwise. Explain your reasoning.

_____

_____

_____

_____

_____

Analyse the relationships between current, voltage and number of turns in each of the two coils of a transformer and apply these ideas to understanding how an electric transformer works

A coil with an AC current input will generate a changing magnetic field. That field will then induce an emf in a second nearby (but electrically separate) coil. This is the basis of the transformer. If the first coil is the primary, P, and the second coil is the secondary, S, then the currents and voltages are related by:

$$\frac{V_P}{V_S} = \frac{N_P}{N_S} \quad \text{and} \quad I_P V_P = I_S V_S$$

where N is the number of turns of each coil.

State numerical answers correct to an appropriate number of significant figures.

1 Explain the use of three transformers that could be found within a typical family home.

_____

_____

_____

_____

_____

_____

_____

_____

_____

_____

_____

_____

_____

_____

_____

_____

_____

_____

_____

_____

2 An ideal transformer consists of two coils of wire wrapped around the same iron core.

a Complete the following table, answering True or False to each statement.

Statement	True	False
i For AC current, such a transformer can be used to give a secondary voltage higher than the primary voltage.		
ii For constant DC current, such a transformer can be used to give a secondary voltage higher than the primary voltage.		
iii When the secondary voltage is lower than the primary voltage, the secondary current is lower than the primary current.		
iv When the primary coil has more turns than the secondary coil, the primary coil has a higher current than the secondary coil.		
v A step-up transformer has more turns on the secondary coil than the primary coil.		

b Provide a sentence of justification for each answer in a.

i _____

_____

ii _____

_____

iii _____

_____

iv _____

_____

v _____

_____

3 A particular model of electric bar heater cannot use the mains power – it is designed to run on a different voltage – and so a transformer is used to run it. Assume it is an ideal transformer. The secondary coil has 500 turns of wire. The primary coil has a current of 6.1 A. The heater uses energy at a rate of 1.4 kW and has a resistance of 8.6 Ω.

a Determine the number of turns in the primary coil.

_____

_____

_____

_____

_____

_____

b Calculate the primary coil voltage.

_____

_____

_____

_____

_____

c Deduce whether it is a step-up or step-down transformer.

_____

d The standard voltage used in household power in Australia is 230 V but in the USA it is 110 V. Deduce whether the bar heater was designed for Australia or the USA, and which of the two countries it is now being used in.

_____

_____

_____

7 Applications of the motor effect

WS 7.1 Analysing the properties of DC motors

STUDENT BOOK
Pages 174–181

LEARNING GOALS

Analyse the functions of the components of a simple DC motor

Analyse the production of torque and effects of back emf

We know that a current-carrying wire in a magnetic field experiences a force given by $F = lIB\cos\theta$ (Chapters 5 and 6 in the student book). From mechanics, we know that when a force acts at a distance from the centre of rotation, there is a torque (Chapter 3 in the student book).

The image below shows how these processes can be combined to spin a shaft, and so drive an electric motor.

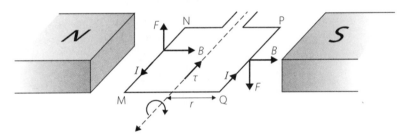

If we use many current loops, we multiply the force and, thus, the torque.

Once the motor is moving, we must switch the direction of the current in the coil to maintain the rotation. The wire that was exerting a force up must now exert a force down, so that the motor will keep turning. This is just like pedalling a bike.

As the coils rotate in a magnetic field, the amount of flux passing through them changes. By Faraday's Law, this generates an emf; by Lenz's Law, it opposes the driving emf. This back emf limits the motor's speed of rotation. The same effect means that a motor can also be a generator – if we spin it using an external force, we can generate an emf. Depending on the wiring, we can then generate AC or DC current.

State numerical answers correct to an appropriate number of significant figures.

1   Use the right-hand rule to deduce the direction of the force on the wire in each of the following diagrams and indicate it with an arrow. In each case, the current in the wire is out of or into the page, as noted.

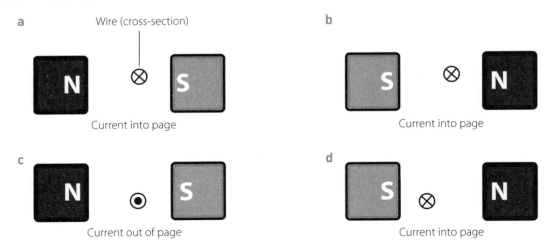

2   The normal of a single loop of wire of cross-sectional area $2.00\,\text{cm}^2$ is at an angle of 90° to a magnetic field of $500\,\text{mT}$. A current of $7.5\,\text{A}$ is flowing in the wire.

   a   Calculate the torque on the loop.

   _____

   _____

   _____

   b   127 loops of the same size, all carrying the same current, are added to the first. What is the torque now?

   _____

   c   At what angle does the torque drop to 15% of its maximum value?

   _____

   _____

   _____

3   Explain why most real DC motors use several coils that are offset from each other, as in the image below, rather than a single coil.

Dreamstime.com/Fireflyphoto

   _____

   _____

   _____

   _____

   _____

   _____

   _____

   _____

   _____

4 The table lists six components of typical DC motors, followed by a list of descriptions, labelled A to F.

   a Put the letter for the matching definition in the box beside the component name.

commutator	
permanent magnets	
carbon brushes	
coils of wire	
armature	
casing	

   A contains all the components

   B reverses the direction of current in the coils as the motor turns

   C holds the coils

   D supply the magnetic field that drives the motor through interaction with the field induced by the current

   E generate a magnetic field when the current passes through them

   F allow current to flow from the source into the rotating coils

   b Which of these six components are usually found as part of the rotor?

   _____

   c Which of these six components are usually found as part of the stator?

   _____

Analyse the operation of simple DC motors, AC induction motors and AC and DC generators

If a voltage or emf is supplied across the terminals of a motor, the motor will experience a torque and turn. If we turn a motor (by hand or using a machine), an emf can be generated across the terminals. A generator is a motor in reverse. When a motor coil moves through the field as a result of the supply emf, the coil will experience a change in flux. This change in flux will induce an emf in the coil (by Faraday's Law) that will oppose the supply emf (by Lenz's Law) and, therefore, reduce the emf and, consequently, the current, through the coil.

An AC induction motor uses a time-varying magnetic field (caused by the changing current in stator coils) to induce a current in a conductor (called a squirrel cage) that is free to rotate. The field generated by the induced current in the squirrel cage then interacts with the field generated by the stator coil, generating a force as the two fields interact and, thus, a torque results, spinning the rotor.

Most large motors driven directly by mains power (for example in a washing machine or an electric pump) are AC induction motors.

State numerical answers correct to an appropriate number of significant figures.

1   By Faraday's and Lenz's Laws, if a loop of wire is mechanically rotated in a magnetic field, the changing flux through the loop will induce a current in the wire. This is the basis of most electricity generation (the main exception is photovoltaic solar power).

Compare the back emf of a motor and the emf generated by a simple generator.

2   A generator produces AC current at 60 Hz and with a maximum voltage of 700 V. It starts with the normal to the plane of the coil perpendicular to the field.

a   What is the period of rotation of the generator?

_____

_____

b   Sketch a graph of the emf of the generator with time, for the first 0.03 s.

c   Qualitatively, describe what would happen to the graph in b if the following changes are made.

i   The number of turns in the coil is decreased.

_____

_____

ii The area of the coil is increased.

_____

_____

_____

iii The magnetic field is increased.

_____

_____

iv The generator spins faster.

_____

_____

d The conducting slip rings are replaced with a split-ring commutator, turning it into a DC generator. Sketch the emf vs time graph.

_____

_____

Exploring relationships between conservation of energy, Lenz's Law, back emf and magnetic braking

STUDENT BOOK
Pages 189–193

LEARNING GOALS

Relate conservation of energy to the operation of electric motors and generators

Lenz's Law is a consequence of the conservation of energy. The fact that a changing magnetic field causes a current to flow is the basis of magnetic braking. If the relative movement of a magnet near a conductor is creating a current, the energy to induce the current must come from somewhere. It comes from the energy of motion, and braking occurs due to conservation of energy. Further, Lenz's Law says that the currents in the conductor will set up a field to oppose the change in flux, which also opposes the relative motion of the magnet and the conductor. This can be used to explain that braking effect. Back emf can be seen as a consequence of the same process.

State numerical answers correct to an appropriate number of significant figures.

1   A DC motor is switched on at time $t = 0\,$s.

    a   Describe the circumstances in which the current through the motor would be a maximum.

    _____

    _____

    b   Describe the circumstances in which the current through the motor would be a minimum.

    _____

    c   If the motor is able to spin freely with only small friction or a small load, sketch a possible qualitative graph of the current as a function of time.

    Consider the relative current values when the motor is yet to start moving ($t = 0\,$s), the period during which the motor speeds up and when the motor has reached its maximum rate of rotation.

2 Consider the following diagram.

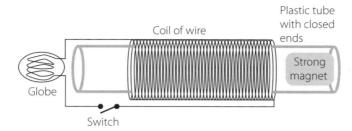

The schematic diagram illustrates a torch, with a coil of wire in a circuit with a globe and a cylindrical magnet that can move. The torch can be illuminated by moving it left and right so that the strong magnet moves through the coil.

a Why does the left–right motion result in the torch illuminating (assuming the switch is closed)?

_____

_____

_____

b Compare the motion of the magnet when the switch is closed with when it is open. Assume that the torch is moved the same way each time.

_____

_____

_____

_____

_____

c The device might be improved by placing a rechargeable battery in the circuit. How might this improve the performance of the torch?

_____

_____

_____

_____

3 a Exercise bicycles in gyms use electromagnetic braking as an effective way of adjusting the level of 'resistance' that the bicycle offers the cyclist. The wheel spun by pedalling is made of an aluminium disc that spins between a pair of magnets, with opposing poles facing each other, as illustrated in the diagram below. The position of the magnets can be adjusted so that they move to cover a greater area of the aluminium disc. Explain how this can make it harder to pedal the bicycle.

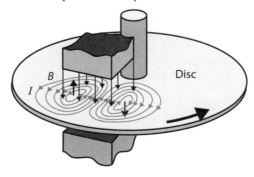

_____

_____

_____

_____

_____

_____

b The creation of eddy currents in the base of a saucepan can be used in cooking. This is the basis of an induction cooktop. Explain how the cooktop could create eddy currents in the saucepan.

_____

_____

_____

_____

4 Inductive (wireless) battery chargers are available for a variety of devices such as electric toothbrushes. Describe what would be required in the battery charger so that an electric toothbrush can be charged by inductive changing.

_____

_____

_____

_____

_____

5 Explain why conservation of energy implies an electric motor will generate a back emf.

_____

_____

_____

_____

_____

Module six: Checking understanding

Circle the correct answer for questions 1–5.

1 A particle of charge $5.0 \times 10^{-9}\,\text{C}$ experiences an acceleration of $0.38\,\text{m s}^{-2}$ in an electric field of $26\,\text{kV m}^{-1}$. What is its mass?

    A $3.4\,\text{kg}$

    B $3.4\,\text{g}$

    C $3.4\,\text{mg}$

    D $0.34\,\text{g}$

2 When a moving charged particle enters a magnetic field, the field does no work on the particle. This is because:

    A   a magnetic field cannot cause a charged particle to accelerate.

    B   the force on the moving particle is perpendicular to the velocity.

    C   the particle must be moving in order for a force to arise.

    D   the particle will not experience a force.

3 The force between two current-carrying wires:

    A   increases as the wires are brought closer together.

    B   decreases when the current in one wire decreases.

    C   is the same whether considering the force of wire 1 on wire 2 or of wire 2 on wire 1.

    D   All of the above.

4 Which of the following statements is not correct?

    A   The induced emf in a wire connecting any two points in a changing magnetic field is independent of the path of the wire.

    B   Electromagnetic induction occurs when the magnetic flux through an area changes with time.

    C   If a loop of wire is placed in a magnetic field and the flux through the loop changes, an emf will be induced across the two ends of the loop.

    D   emf is the energy per unit charge available to a charged particle.

5 The component of a DC motor that switches the current direction when the motor rotates is:

    A   a split ring commutator.

    B   a pair of slip rings.

    C   a pair of carbon brushes.

    D   the armature.

6 The force per unit length between parallel current-carrying wires is given by:

$$\frac{F}{l} = \frac{\mu_0 I_1 I_2}{2\pi r}\text{, where } \mu_0 = 4\pi \times 10^{-7}\,\text{T m A}^{-1}.$$

If the current in wire 2 is triple that in wire 1, what is the current in wire 1 given that the force per unit length is $6.3 \times 10^{-5}\,\text{N m}^{-1}$ when the wires are $1.2\,\text{cm}$ apart?

7 Faraday's Law for the emf induced by a changing magnetic field can be written: $\varepsilon = -\dfrac{\Delta \Phi}{\Delta t} = \dfrac{-\Delta(BA\cos\theta)}{\Delta t}$.

   a What quantity is represented by each of the following symbols?

   ε: _____

   Φ: _____

   Δ: _____

   b From which law does the negative sign come? State that law briefly.

   _____

   _____

Reviewing prior knowledge

State numerical answers correct to an appropriate number of significant figures.

1 Distinguish between the two main types of wave based on direction of oscillation relative to direction of propagation.

_____

_____

2 Match the description to the wave behaviour.

A	Reflection	1	When a wave front changes direction or bends when passing an opening.
B	Refraction	2	The change in direction of a wavefront when two surfaces of different medium meet, such that the wavefront returns into the medium in which it originated.
C	Diffraction	3	When two or more waves interact, causing either constructive or destructive interference.
D	Superposition	4	When a wave front changes direction when passing from one medium to another.

3 If a radio wave has a frequency of 6.0×10^8 Hz, calculate:

a its wavelength.

_____

b its period.

_____

4 When light travels from one medium into another, it changes its speed depending on the material that it is travelling through. Describe the mathematical relationships that describe this phenomenon.

_____

_____

5 What frequency is observed and what is the frequency change when an ambulance is approaching a stationary pedestrian at $34\,\mathrm{m\,s^{-1}}$ with the frequency of the siren at $1230\,\mathrm{Hz}$? Take the speed of sound in air as $340\,\mathrm{m\,s^{-1}}$.

_____

_____

8 Electromagnetic spectrum

WS 8.1 Investigating Maxwell and the electromagnetic spectrum

STUDENT BOOK
Pages 203–205

LEARNING GOALS

Investigate Maxwell's contribution to the unification of electricity and magnetism, including the prediction of electromagnetic waves and their velocity

Describe how science is moved forward through making and testing predictions

Many great scientists and philosophers throughout history thought light acted through a medium, as is the case with sound, and they hypothesised the existence of a medium that covered everything in the Universe. This hypothetical medium was called several different names, including the *plenum* and the *aether*. It was not until 1865 that James Clerk Maxwell linked electricity to magnetism and predicted that this unification, or electromagnetic wave, was light.

1   By doing a small amount of research, complete the box below on the history of light. (This is not part of the syllabus but gives context to the work done by Maxwell.)

WHO	WHAT THEY THOUGHT	WHEN
Pythagoras		5th century BCE
Euclid		3rd century BCE
Hasan Ibn al-Haytham		1011–1021
René Descartes		1629–1633
Isaac Newton		1672
Christiaan Huygens		1678

2   Describe the meaning of each of the four equations that Maxwell used to form his theory of electromagnetism.

a   Gauss' Law of Electricity

_____

_____

b   Gauss's Law of Magnetism

_____

_____

c   Faraday's Law of Induction

_____

_____

d   Ampere's Law (modified)

_____

_____

3   a   Explain why Maxwell predicted that light was an electromagnetic wave.

_____

_____

_____

   b   How did Maxwell's work allow him to predict that light was a part of a spectrum and not just the light we can see?

_____

_____

_____

4   Maxwell died before his predictions were proved correct through experimentation. Why is it important in science to make and test predictions?

_____

_____

_____

Describe how electromagnetic waves are produced and how this relates to Maxwell's predictions

Identify predictions made by Maxwell about electromagnetic waves

1 **a** Below is a diagram of an electromagnetic wave. Identify what is represented by the lines labelled E and B.

E: _____

B: _____

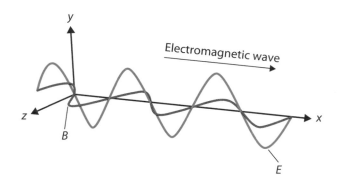

b Outline how an electromagnetic wave is produced.

_____

_____

_____

2 Outline key predictions made by Maxwell about electromagnetic waves.

_____

_____

_____

_____

_____

_____

_____

_____

_____

_____

3 Radio waves, microwaves, infra-red radiation, visible light, ultraviolet radiation, X-rays and Gamma rays are all examples of classifications of electromagnetic waves. Choose one to illustrate how Maxwell's prediction of electromagnetic wave production was correct.

_____

_____

_____

_____

_____

_____

WS 8.3 Investigating the speed of light

STUDENT BOOK
Pages 207–214

LEARNING GOALS

Investigate the experimental evidence of the measurement of the speed of light over the years

Analyse an experiment that has been carried out to measure the speed of light

1 Do some brief research to complete the table from the information provided, matching the correct scientist to the experiment, the measurement and the year that they carried out the work.

Scientists: Michelson, Foucault, Galileo, Romer, Froome, Bradley, Fizeau, Weber and Kohlrausch, Woods

Years: 1676, 1728, 1849, 1856, 1862, 1926, 1958, 1978, 1638

Experiments: Lanterns on hills, Rotating mirror, Toothed wheel, Wavelengths and frequencies of lasers, Rotating mirror, Eclipse of the moon Io, Radio interferometry, Stellar aberration, Measurements of μ_0 and ε_0

Measurement in $km\,s^{-1}$: 298 000, 220 000, Instantaneous, 301 000, 310 000, 299 796 ± 4, 299 792.5 ± 0.1, 315 000, 299 792.459 ± 0.001

SCIENTIST	YEAR	EXPERIMENT	MEASUREMENT IN $km\,s^{-1}$
	1638		
	1676		
	1728		
	1849		
	1862		
	1909		
	1926		
	1958		
	1978		

2 Evaluate the development of the measurement of the speed of light, from one earlier attempt to the accepted value of today.

HINT

Make sure that you include any relevant information regarding factors such as improving technologies for more accurate results.

_____

_____

_____

_____

_____

_____

_____

_____

_____

_____

_____

_____

_____

_____

3 Fong set up an experiment to measure the speed of light using a microwave. He spread marshmallows on a plate, which he then put into the microwave on full power for 30 seconds. He checked the marshmallows for melted spots and continued until he saw definite hot (melted) spots in the marshmallows. He then measured the distance between the spots and used this information to work out the speed of light.

a Explain how Fong would be able to measure the speed of light from this experiment.

_____

_____

_____

_____

_____

_____

b Identify the problems with this experiment from the description.

_____

_____

_____

_____

LEARNING GOALS

Describe how light causes spectra through interactions with gaseous atoms

Analyse and differentiate between spectra

Explain how spectra can be used to gain information about stellar bodies

When an electron in a gaseous atom interacts with a photon, it will gain energy and become excited and change orbit around the nucleus. When the electron releases this energy in the form of another photon, it becomes deexcited and returns to its ground state.

1 Name the three types of light spectra. Describe how each is formed and its appearance.

_____

_____

_____

_____

_____

_____

_____

_____

2 Every element emits or absorbs specific wavelengths of light. Using this information, you can identify the elements that are in a given sample. Below are the emission spectra for four known elements. Use them to identify the elements that are in the samples.

Hydrogen

Helium

Lithium

Beryllium

Wavelength (nm)

a Sample 1: _____

b Sample 2: _____

c Sample 3: _____

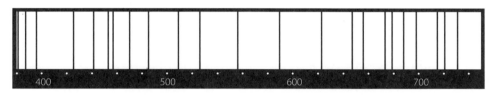

d Sample 4: _____

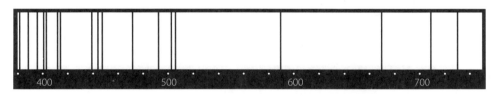

e Sample 5: _____

f Sample 6: _____

Wavelength (nm)

3 Our understanding that elements give out unique spectra allows us to discover the chemical composition of stars. It also allows us to understand other properties of stars.

a Identify the property that could be determined using this graph and justify your answer.

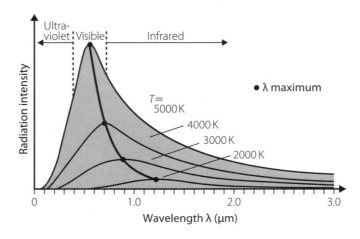

_____

_____

_____

b The centre spectrum in the diagram below is that of Element X as seen on Earth. Explain what information you could gain from the other two spectra shown.

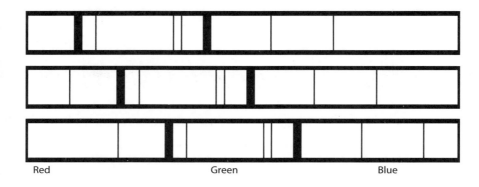

_____

_____

_____

_____

_____

c Explain how stellar spectra can be used to identify the size of a star.

_____

_____

_____

_____

_____

WS **9.1** ## Analysing the experimental evidence of Newton and Huygens

STUDENT BOOK
Pages 227–231

LEARNING GOALS

Understand how different theories throughout history have tried to explain the movement of light

Analyse evidence from competing models and see how they are used to explain natural phenomena

The question of whether light is a wave or a particle has been investigated throughout history. Several models have been suggested and, as with most models, there are parts that have been shown to be correct and parts that have been shown to be incorrect. Christiaan Huygens and Isaac Newton proposed competing models.

1 Describe Newton's model of light.

_____

_____

_____

2 Describe Huygens' model of light, and his resulting principle.

_____

_____

_____

_____

_____

3 Create a table to compare how the two models explained the phenomena of reflection, refraction, diffraction and rectilinear propagation.

4 Analyse the ability of each model to explain light behaviours and explain why Newton's model was more readily accepted by the scientific establishment than Huygens'.

_____

_____

_____

_____

_____

_____

_____

_____

_____

_____

_____

_____

_____

_____

_____

_____

_____

_____

_____

5 For each wave propagating below, draw the new wavefront according to Huygens' principle.

a

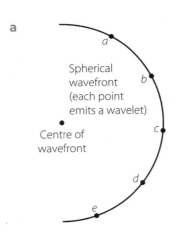

b

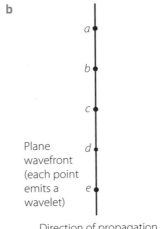

Direction of propagation

WS 9.2 Analysing interference and diffraction of light

STUDENT BOOK
Pages 232–242

Identify the conditions required for light to cause interference patterns

Compare the resulting interference patterns due to the conditions used (i.e. single slit/double slit/diffraction grating)

Understand, both qualitatively and quantitively, that light can be diffracted and have interference patterns suggested by the classical wave model

If light is a wave it will exhibit wave behaviours. Consequently, it would be expected that light will diffract when it encounters an object or openings in a barrier. Interference would result from this diffraction and result in a characteristic pattern.

State numerical answers correct to an appropriate number of significant figures.

1 'Young's double slit experiment definitively proved that light is a wave.' Analyse this statement using your knowledge of the experiment and the behaviour of light.

2 Draw a simplified set-up of Young's experiment below, showing the wave pattern as well as the regions of constructive and destructive interference.

3 Identify the two conditions for the light to make the double-slit experiment successful, and explain why they are needed.

_____

_____

_____

_____

_____

4 A student sets up an investigation into double-slit diffraction as in the image below. In the space below draw what you would expect to see on the screen and explain your answer. In this experiment $d = 0.05\,\text{mm}$, $\lambda = 696\,\text{nm}$ and $L = 2\,\text{m}$.

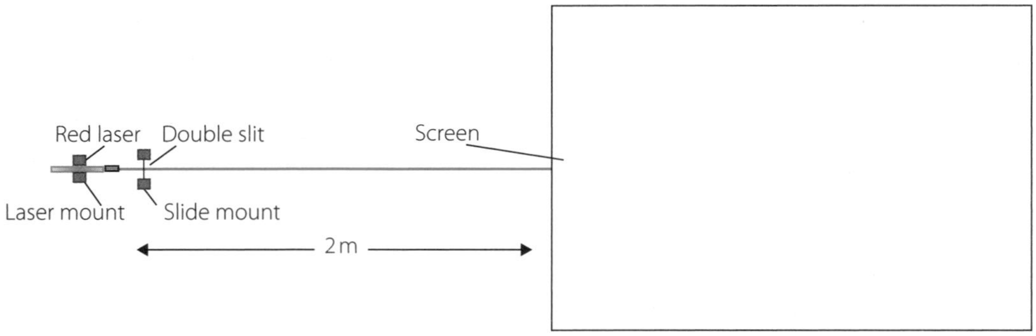

Red laser   Double slit   Screen

Laser mount   Slide mount

— 2m —

_____

_____

_____

_____

5 a In the above investigation, what would be the distance between the central maximum and the first maximum?

_____

_____

b What would be the angle to the first maxima?

_____

_____

6 When a bright direct light strikes a circular disc at 90 degrees a shadow will be cast on a screen parallel to the disc. In the centre of the shadow a bright spot can appear – this is called Poisson's spot. Explain, using a diagram, from your knowledge of diffraction, how Poisson's spot is formed.

----

7 Some students were setting up a diffraction grating experiment to observe the wave behaviour of light. They used an uncovered incandescent light bulb as their light source and a diffraction grating with 500 slits per mm. They used a ruler to measure the distance between the diffraction grating and the wall.

a Create a risk table listing some of the risks associated with this experiment, the risk level and how the students might mitigate the risk.

b Would this experiment be the most valid way to test for the diffraction of light? If not, how would you improve the method? Justify your answer.

----

8 **a** Compare the different interference patterns that you would expect to see from single-slit interference, double-slit interference and diffraction gratings.

_____

_____

_____

_____

_____

_____

_____

b Label each of the three patterns seen in the diagram below as one of: single-slit, double-slit or diffraction grating.

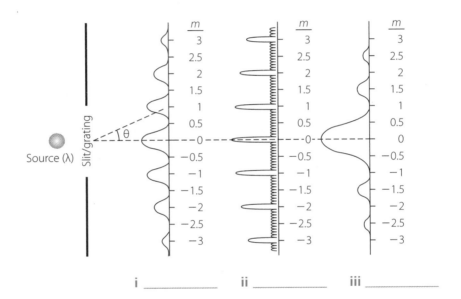

i _____ ii _____ iii _____

Analysing experimental evidence for the polarisation of light

Understand that light can be polarised due to wave-like properties and that this can be analysed quantitatively using Malus' Law

Analyse experimental evidence for the polarisation of light according to Malus' Law

The wave model of light states that light is formed by perpendicular oscillating electric and magnetic fields that induce each other and act at right angles to each other. As this can happen in a 360° field, it would stand to reason that if light is a wave then it could be polarised, while if it is a particle it could not be polarised.

State numerical answers correct to an appropriate number of significant figures.

1 Explain how the intensity of light changes when it passes through a single polarising filter.

_____

_____

2 A group of students is planning to test Malus' Law. The students aim to investigate what happens if they move the second polariser through 180° relative to the first polariser.

They have an unpolarised light source ($500\,\mathrm{W\,m^{-2}}$), a set of two polarising filters (a polariser and an analyser) and a light meter.

a Write a suitable method for this aim.

_____

_____

_____

_____

_____

b The students record the following data for the investigation. Use Malus' Law to complete the table.

Angle between filter 1 and filter 2 (°)	Light intensity through the first filter ($\mathrm{W\,m^{-2}}$)	Light intensity through the second filter ($\mathrm{W\,m^{-2}}$)
0		
10		
20		
30		
40		
50		
60		
70		
80		
90		
100		
110		
120		
130		
140		
150		
160		
170		
180		

c Plot a graph of the data below.

d Use the shape of the plotted graph to analyse the results.

_____

_____

_____

e Was this a valid and reliable investigation?

_____

_____

_____

_____

3 An unpolarised light source of $3000\,\mathrm{W\,m^{-2}}$ is shone onto two polarising filters, which are aligned so that no light comes through them. A third filter is placed in between these filters so that the plane of polarisation is at an angle of 45° to them both.

a What would the angle between the planes of polarisation of any two adjacent filters need to be to ensure no light passes through?

_____

9780170449687

b Calculate the intensity of the light transmitted when the third filter is added.

_____

_____

_____

_____

4 Calculate the angle between two polarisers when the transmitted light intensity is 42% of the original light intensity.

_____

_____

_____

_____

_____

_____

5 Calculate the percentage increase in the intensity of light seen when the analyser is rotated by 15° and the initial angle between the polariser and the polarising axis is 45° (noting that the angle will decrease after rotation).

_____

_____

10 Light: quantum model

WS 10.1 Understanding the contributions of Max Planck and black body radiation to the development of the quantum model of light

STUDENT BOOK
Pages 251–258

LEARNING GOALS

Analyse evidence of black bodies and how this led to Planck's idea of the quantisation of light

Assess Wien's Law and this contribution to the understanding of black body radiation

Apply Wien's Law to analyse celestial temperatures

State numerical answers correct to an appropriate number of significant figures.

1 Describe what is meant by black body radiation and the characteristics of a black body radiator in relation to emission and absorption of radiation.

_____

_____

_____

2 Explain the ultraviolet catastrophe.

_____

_____

_____

_____

_____

_____

_____

_____

3 a Use Wien's Law to calculate the peak wavelength (wavelength of maximum intensity) for the star Alpha Crucis, which has a surface temperature of 30 000 K.

_____

b If the peak wavelength for Betelgeuse is 828 nm, calculate its temperature.

_____

4 Explain the assumptions Planck made that enabled him to explain the black body curve seen experimentally.

_____

_____

_____

_____

5 Planck showed that the energy for a quantum of light could be calculated from the relationship $E = hf$. He called this an 'act of desperation' and a mathematical trick.

   Why would a scientist say this about a solution that they had found, and how might it affect their investigations?

_____

_____

_____

_____

_____

_____

_____

6 Calculate the energy of a photon that has a frequency of 8×10^{14} Hz.

_____

_____

7 What is the frequency of a photon with energy 4.3×10^{-18} J?

_____

8 Because the energy of a photon is very small, the electron volt (eV) is often used as the unit of energy. This is the amount of energy gained by an electron accelerating through a potential difference of 1 volt and is equal to 1.602×10^{-19} J. Calculate the energy in eV of a photon with a frequency of 7.6×10^{16} Hz.

_____

_____

Analyse the evidence for the photoelectric effect

Graphically show the threshold frequency and work function for a metal

Evaluate significant contributions to the understanding of the photoelectric effect

The first evidence for the photoelectric effect was seen in the late 1800s, but the experimental evidence was not understood until Einstein used the notion of quantisation of energy, proposed by Planck, to suggest that light might not only be a wave as previously thought.

State numerical answers correct to an appropriate number of significant figures.

1 Research the major developments that contributed to the understanding of the photoelectric effect.

_____

_____

_____

_____

_____

_____

_____

2 Explain what is meant by:

a work function.

_____

_____

_____

b threshold frequency.

_____

_____

3 Complete the table below contrasting the theoretical expectations from the wave model of light and the observed experimental results.

Theoretical prediction using the wave model of light	Experimental evidence
Electrons should be emitted at all frequencies of light.	
At low intensities there should be a delay between the light hitting the metal surface and electrons being emitted.	
The current produced should be dependent on both the intensity and the frequency.	
Maximum kinetic energy should be related to the intensity of the light and not the frequency.	

4 Identify and explain the function of the parts of a photoelectric effect apparatus in the image below. Choose your labels from the following list.

Anode   Incident light   Cathode
Vacuum chamber   Photoelectrons   Ammeter
Voltmeter   Power source   Variable resistor

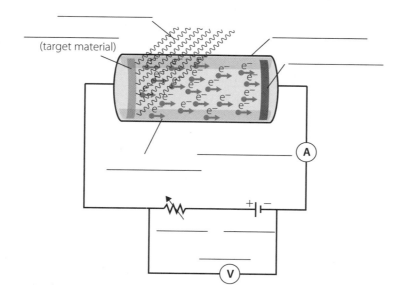

(target material)

_____

_____

_____

_____

_____

_____

_____

_____

_____

_____

Questions **5** to **12** relate to an investigation that was carried out to find out the threshold frequency for a piece of metal. The aim of the investigation was to use the stopping voltage to calculate the maximum kinetic energy of an emitted electron to find the threshold frequency.

5 **a** The results are as below. Complete the missing columns that would be required for this investigation.

Wavelength (nm)	Stopping voltage (V)	Energy (eV)	Frequency (Hz)
166	−4.49		
187	−3.87		
214	−3.00		
250	−2.12		
300	−1.37		

b Graph the results below.

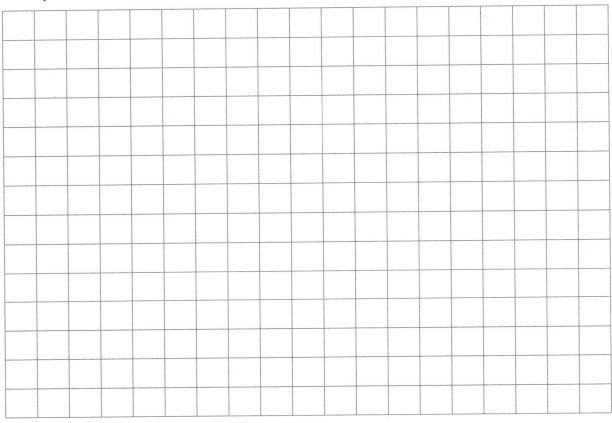

c From your graph, identify the threshold frequency of the metal in the investigation.

_____

6 What is the work function of the metal in both J and eV?

_____

_____

7 Calculate the kinetic energy of the fastest moving electron from the results of this investigation.

_____

_____

_____

8 At what speed is this electron moving?

_____

_____

9 Compare the maximum speeds of the emitted electrons when the metal in this experiment is exposed to light of the longest and shortest wavelengths.

_____

_____

_____

_____

_____

10 Calculate the gradient of the slope and show how you could use this to comment on the results of the investigation.

_____

_____

_____

11 Propose what you would expect to find if the experiment was performed with other metal surfaces exposed to the same incident light.

_____

_____

_____

12 Write a suitable conclusion for the results of this investigation.

_____

_____

_____

Questions **13** to **15** relate to an investigation carried out by a group of students to determine the effects of changing the intensity of light on the current produced by the photoelectric effect.

13 **a** Write a suitable hypothesis for this investigation.

_____

Photoelectric effect apparatus containing different size apertures and different colour filters was used for this experiment.

The method was:

1 Record the diameter of each of the four apertures.

2 Calculate the area of each aperture.

3 Set up the apparatus with the smallest wavelength filter (428 nm).

4 Backing voltage should be adjusted until it reads zero.

5 Measure the current produced.

6 Change the colour filter.

7 Repeat the steps 3–6 for the remaining apertures and colours.

8 Record your data in a table.

b Identify whether the method above would give a valid answer to the hypothesis. Justify your answer.

_____

_____

c Suggest an additional change the investigators need to make for a valid investigation.

_____

14 a Complete the table of results below and plot a graph of the completed data.

Diameter of aperture (mm)	Area of aperture (mm^2)	Current (mA)
7		0.10
10		0.17
14		0.22
20		0.26

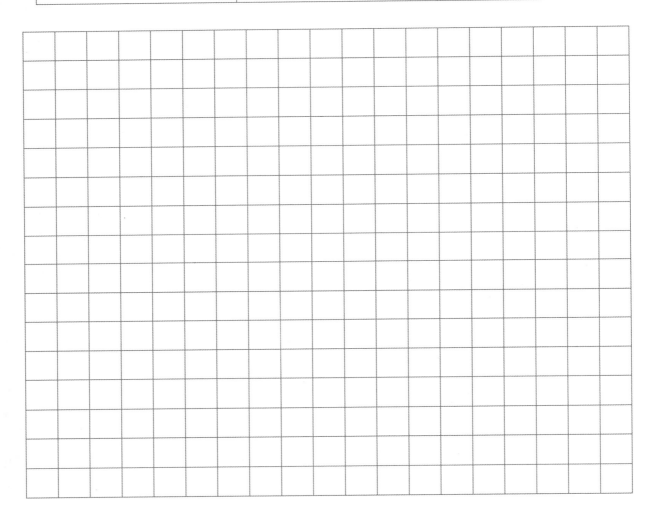

9780170449687

b Analyse the results in your graph.

_____

_____

_____

_____

15 Write a suitable conclusion to this investigation.

_____

_____

16 Show that the photoelectric effect as described by the equation $K_{max} = hf -$ work function follows the law of conservation of energy.

_____

_____

_____

WS 11.1 Investigating reference frames and Einstein's postulates

STUDENT BOOK
Pages 273–281

LEARNING GOALS

Identify the difference between inertial and non-inertial frames of reference

Understand the postulates that underpin special relativity

Analyse and evaluate the evidence supporting Einstein's postulates

1 A ball is dropped from the top of the mast of a moving boat. Draw the predicted path of a ball when it is dropped from the top of the mast from the frames of reference of someone on the boat and someone watching from shore. Use diagram **a** for the reference frame of the boat and diagram **b** for the frame of reference for the shore.

a

b

2 Explain the difference between the two paths in Question **1**.

_____

_____

_____

_____

3 Which frame of reference is correct? Justify your answer.

_____

_____

_____

4 **a** According to classical physics, how can you tell if you are in a moving or stationary frame of reference?

_____

_____

b What could you deduce from this explanation?

_____

_____

_____

5 According to classical relativity, there is no absolute frame of reference. However, thinking at the time was that light required a medium to travel through, a medium they called the aether. Suggest why the aether would present a problem to classical relativity.

_____

_____

_____

_____

6 a Identify and describe, with the help of a labelled diagram, a famous experiment that showed there was no evidence of an aether.

_____

_____

b What consequence did this have for our understanding of how light behaves?

_____

_____

7 Analyse this experiment in terms of its reliability and validity.

_____

_____

_____

_____

_____

8 State Einstein's two postulates for special relativity.

_____

_____

_____

9 a In an experiment on atomic decay, scientists were analysing subatomic particles that were accelerated to close to the speed of light and at the same time were emitting gamma rays. Using Galilean or classical relative motion laws, predict how fast the emitted gamma rays were travelling.

_____

_____

_____

b Why would this prediction not be borne out?

_____

_____

_____

10 List all the current evidence against Einstein's postulates, and discuss what this means for the understanding of special relativity.

_____

_____

_____

Investigating time dilation and length contraction

Identify what is an appropriate thought experiment

Apply the formula for time dilation and length contraction

Quantitatively analyse the evidence for both time dilation and length contraction

After Einstein put forward his Theory of Special Relativity, the way that people viewed the world was changed. It turned quantities that were definite, such as time and length, into quantities that were relative to the observer.

State numerical answers correct to an appropriate number of significant figures.

1 Why was the international standard of the metre changed from the length of a metal bar to the distance travelled by light in a certain time period?

_____

_____

_____

_____

_____

2 Einstein was famous for using thought experiments to predict how things will behave when they reach relativistic speeds. Define a thought experiment and outline why they are a valid way to investigate the idea of special relativity.

_____

_____

_____

_____

_____

3 Outline the difference between a thought experiment and a paradox.

_____

_____

_____

4 Einstein devised a thought experiment regarding a light on a train to help understand time dilation. Using diagrams, show your understanding of this experiment.

_____

_____

5 Identify the apparent paradox regarding time dilation and discuss how it can be resolved.

_____

_____

_____

_____

_____

_____

_____

For questions **6–10** we will use the time dilation formula $t = \dfrac{t_0}{\sqrt{1 - \dfrac{v^2}{c^2}}}$.

6 An astronaut on a spaceship travelling past Earth at $0.75c$ uses his stopwatch to record a 30 s period for an event occurring on the spaceship. How long would an observer on Earth record for the same event on the spaceship? How long would an observer on Earth record for the same period?

_____

_____

_____

7 Eric is waiting for his friend Heather to arrive from a nearby galaxy. Eric has waited for 3 hours when Heather arrives. Heather's spaceship was travelling at $0.82c$. What time period has elapsed on the spaceship's clocks?

_____

_____

8 Leo waited 10 years for his sister Myra to complete a journey to the Andromeda galaxy. Myra measured that she was gone for 6 years. How fast was Myra travelling?

_____

_____

_____

9 Complete the table for the Lorentz factor when travelling at different speeds, then plot a graph of γ against $\dfrac{v}{c}$.

v (m s^{-1})	$\dfrac{v}{c}$	γ
3.00×10^2		
3.00×10^4		
3.00×10^7		
1.40×10^8		
2.00×10^8		
2.50×10^8		
2.90×10^8		
2.99×10^8		
2.999×10^8		

10 What does this graph indicate in relation to the effects of time dilation when approaching the speed of light?

_____

_____

The previous questions dealt with time dilation and thought experiments. The next part of this worksheet focuses on length contraction. Any quantitative questions will use the length contraction formula: $L = L_0\sqrt{1-\dfrac{v^2}{c^2}}$.

11 Propose why length contraction is a logical consequence of the speed of light being constant.

_____

_____

_____

_____

12 A train, when stationary, is measured as having a length exactly equal to the length of a tunnel it is to travel through. When the middle of the moving train is exactly in the middle of the tunnel, how will a passenger travelling on the train view it compared to a stationary observer outside the tunnel?

_____

_____

_____

_____

13 A company is planning to send a crew to Mars. They are using new technology that can get their spaceship to travel at 0.65c. The distance between Earth and Mars is approximately 113.39 million km. How far would the crew in the spaceship perceive this to be?

_____

_____

14 How fast would an 8 m long race car need to be travelling to appear 6.5 m long to a stationary observer?

_____

_____

_____

15 Bill is traveling to Betelgeuse at 0.65c, which causes his perception of the distance travelled to be 532 light years. How far would Aissa on Earth measure this distance to be?

_____

_____

16 Explain how the detection of muons at Earth's surface is evidence of time dilation.

_____

_____

_____

_____

_____

_____

_____

_____

17 a The Hafele–Keating experiment showed evidence for time dilation using atomic clocks. They used four caesium atomic clocks and placed two at the United States Naval Observatory in Washington DC and two on commercial planes flying eastwards and westwards, respectively. What was the purpose for the positioning of the clocks?

_____

_____

_____

_____

_____

The difference between the ground-based clocks and those on the respective planes is recorded below. (Note these results are a consequence of special relativity and general relativity, which is not part of the Stage 6 course.)

	EASTWARD JOURNEY (ns)	WESTWARD JOURNEY (ns)
Predicted	−40 ± 23	+275 ± 21
Observed	−59 ± 10	+273 ± 7

b How did these results show evidence for time dilation?

_____

_____

Applying relativistic momentum and mass–energy equivalence

LEARNING GOALS

Understand that relativistic momentum imparts a universal speed limit on objects with mass

Apply the equation $E = mc^2$ to show how energy can be released from matter

The first postulate of special relativity states that the laws of physics hold true in all inertial frames of reference. Does this mean that momentum is conserved in relativistic frames of reference? If momentum is conserved, what impact does this have on the kinetic energy of an object?

State numerical answers correct to an appropriate number of significant figures.

1 Define rest mass.

_____

_____

2 An object of 1 kg mass travels at different relativistic velocities. Complete the following table of relativistic momentum and Newtonian momentum (classical momentum), and plot the data in a graph.

VELOCITY $(m\,s^{-1})$	NEWTONIAN MOMENTUM $(kg\,m\,s^{-1})$	RELATIVISTIC MOMENTUM $(kg\,m\,s^{-1})$
3.0000×10^2		
3.0000×10^5		
3.0000×10^7		
1.5000×10^8		
2.0000×10^8		
2.5000×10^8		
2.9000×10^8		
2.9995×10^8		
2.9999×10^8		

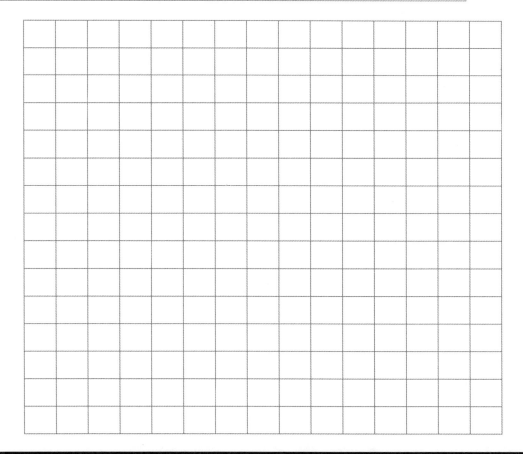

3 The momentum of an object changes when it approaches the speed of light, therefore the energy of the object must also change. Propose why this is the case.

_____

_____

_____

4 What significant implication arose for our understanding of mass after Einstein proposed energy–mass equivalence?

_____

_____

The following problems will use the equation $E = mc^2$.

5 One of the ways the Sun produces energy is through the proton–proton chain. Calculate the amount of energy produced in each stage.

   a $^1_1H + ^1_1H \rightarrow ^2_1H$, given the mass of deuterium (^2_1H) is 3.35×10^{-27} kg and the mass of hydrogen (^1_1H) is 1.67×10^{-27} kg.

_____

_____

_____

_____

   b $^3_2He + ^3_2He \rightarrow ^4_2He + ^1_1H + ^1_1H$, given the mass of ^3_2He is 5.01×10^{-27} kg and ^4_2He is 6.64×10^{-27} kg.

_____

_____

_____

6 When a particle and its antiparticle meet they annihilate each other (both cease to exist) and all the mass of both particles is converted to energy. Calculate the amount of energy released when an electron and a positron annihilate. $m_e = 9.109 \times 10^{-31}$ kg.

_____

_____

7 Compare the percentage of mass converted to energy for each of the processes in questions **6** and **7**.

_____

_____

_____

_____

_____

_____

Module seven: Checking understanding

Circle the correct answer for questions **1–4**.

1 Which of these statements is true regarding Maxwell's understanding of light?

   A That light is an electromagnetic wave, created by oscillating electric and magnetic fields.

   B That visible light is the only type of electromagnetic wave.

   C That electromagnetic waves require a medium to move through.

   D That the speed of light is infinite.

2 A light ray made up of white light is shone onto a slit with a diameter of $50\,\mu m$ with a screen behind it. What will happen to the light?

   A One narrow point of white light will be projected on the screen beyond.

   B The light will diffract with the blue end diffracting the most.

   C The light will diffract with the red end diffracting the most.

   D The light will reflect off the slit, causing no light to go through.

3 What was meant by the ultraviolet catastrophe?

   A When the world will run out of ultraviolet light.

   B Where classical physics failed to explain experimental results for the ultraviolet part of blackbody radiation curves.

   C The photoelectric effect would be difficult to replicate due to it being hard to create ultraviolet rays.

   D That it was too dangerous to work with ultraviolet waves so it was impossible to accurately measure their energy.

4 If a spaceship is travelling at relativistic speeds, past a stationary observer in which direction would the dimensions of the spaceship be observed to contract, and who would observe this?

   A In the direction of travel; the stationary observer

   B In the direction of travel; the person on the ship

   C In the direction perpendicular to direction of travel; the stationary observer

   D In the direction perpendicular to the direction of travel; the person on the ship

State numerical answers correct to an appropriate number of significant figures.

5 Describe an example of relativistic behaviour that could be witnessed on Earth.

_____

_____

_____

6 The graph below shows the line of best fit for data collected, in a photoelectric effect experiment, of energy of emitted photoelectron vs frequency of incident light. Unfortunately, the frequency scale has been omitted, but the equation for the line has been completed.

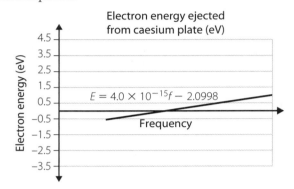

a Calculate the frequency of a photon with a wavelength of 630 nm.

_____

b Using the information in the graph, would light of this frequency eject an electron from the caesium plate?

_____

_____

c A photon with energy of 5.4 eV strikes the same plate. What would be the maximum possible velocity of an ejected electron?

_____

_____

MODULE EIGHT »
FROM THE UNIVERSE TO THE ATOM

Reviewing prior knowledge

1 Identify the three major subatomic particles and describe their features.

2 What is the cause of natural radioactivity?

3 Describe the effect of magnetic and electric fields on charged particles.

4 Explain how the Doppler effect can be used to measure relative motion between a source of waves and an observer.

5 Outline a significant change in the understanding of light that came with the development of quantum physics. (Recall Chapter 10.)

6 Explain the relationship that light has with energy.

7 Explain the implications of Einstein's energy–mass equivalence equation.

Origins of the elements

WS 12.1 Investigating the early Universe's transformation of radiation into matter

STUDENT BOOK
Pages 310–316

LEARNING GOALS

Identify the four fundamental forces

Identify the conditions during the early expansion of the Universe

Explain the effect of these conditions on matter and light.

1 On the diagram below, label the four fundamental forces in the order in which they split from the unified force.

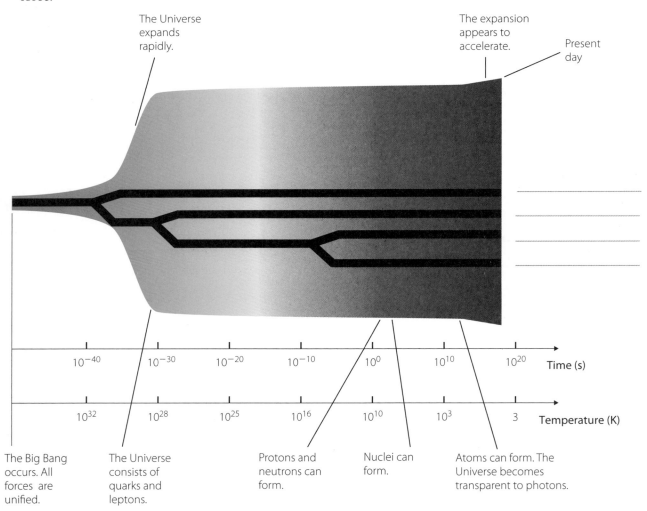

The Universe expands rapidly.

The expansion appears to accelerate.

Present day

Time (s)
10^{-40} 10^{-30} 10^{-20} 10^{-10} 10^{0} 10^{10} 10^{20}

Temperature (K)
10^{32} 10^{28} 10^{25} 10^{16} 10^{10} 10^{3} 3

The Big Bang occurs. All forces are unified.

The Universe consists of quarks and leptons.

Protons and neutrons can form.

Nuclei can form.

Atoms can form. The Universe becomes transparent to photons.

2 Identify the term given to the rapid expansion of the early Universe.

_____

3 Explain why the Universe was opaque before electrons combined with nuclei.

_____

_____

_____

_____

_____

4 The Cosmic Microwave Background Radiation (CMBR) is considered an essential piece of evidence for the Big Bang Theory. It was first discovered in 1965. The following representation of the CMBR indicates the isotropic nature of the early Universe. Lighter regions indicate slightly higher temperatures and darker regions indicate slighter lower temperatures; however, the fluctuations in temperature are incredibly small.

NASA

a Identify what is responsible for the CMBR that we see today.

_____

b Explain why the CMBR is observed in the microwave region of the electromagnetic spectrum.

_____

_____

c What does the relative evenness of temperature across the Universe imply?

_____

_____

_____

_____

_____

_____

5 Describe the evidence supporting the existence of antimatter that was proposed to have been present in the early Universe.

_____

_____

_____

_____

6 Describe the difference between the strong and weak nuclear forces.

_____

_____

_____

_____

Explain how evidence suggested the Universe is expanding

Graph secondary data

Determine values using a line of best fit

For most of history, humans have considered the Universe to be static, neither expanding or contracting. This was a model that was initially supported by Einstein. It was not until the early 20th century that evidence began to indicate an expanding Universe.

Edwin Hubble used analysis of cepheids produced by Henrietta Leavitt to determine the distance to the M31 galaxy (approximately 900 000 light years). He then compared the position of M31 hydrogen spectral lines to hydrogen lines produced in the laboratory (more on stellar spectra in worksheet 12.3). He found that these lines had been Doppler-shifted, indicating that M31 was receding relative to an observer on Earth. This red shift of the spectral lines allowed Hubble to calculate the recession velocity of M31. When he plotted the recession velocity of numerous galaxies against their distance from Earth, he found a positive linear correlation.

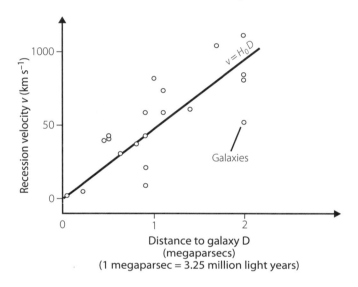

The gradient of this graph became the Hubble constant (H_0), revealing the rate of expansion of the current Universe, and has units of s^{-1}. The inverse of Hubble's constant, therefore, provides an estimate for the age of the Universe.

State numerical answers correct to an appropriate number of significant figures.

1   What limitations existed in our understanding of the size of the Universe before the 20th century?

----

2   Explain why Leavitt's work was so critical to Hubble's discovery.

----

3 With reference to the following diagram, explain the terms 'red shift' and 'blue shift'.

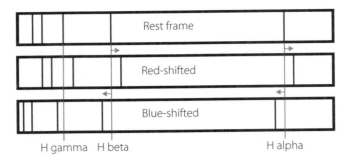

4 Explain how Hubble's discovery supports the Big Bang Theory.

5 The recessional velocity of and distance to a number of extragalactic objects are shown below.

Name	Distance (Mpc)	Recessional velocity (km s^{-1})
Fornax	15	1376
Eridanus	20.7	1617
MDL 59	31.3	2484
Cen 30	43.2	3853
NGC 0507	57.3	4963
NGC 0383	66.6	5325
Cancer	74.3	5962
Coma	85.6	6973
Abell 2634	114.9	8836

a Graph the data to determine Hubble's constant. Note the accepted value for Hubble's constant is $73.8 \pm 2.4 \, \text{km}\,\text{s}^{-1}\,\text{Mpc}^{-1}$.

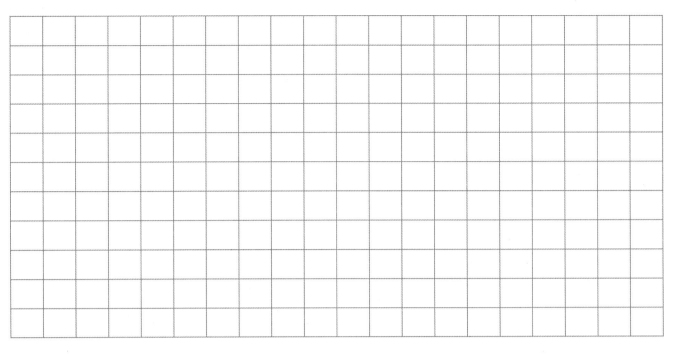

b Using your graph, calculate an estimate of the age of the Universe ($1\,\text{pc} = 3.085 \times 10^{16}\,\text{m}$).

_____

_____

6 Comment on the accuracy, reliability and validity of this determination.

_____

_____

_____

_____

_____

_____

_____

_____

_____

_____

_____

_____

_____

_____

LEARNING GOALS

Explain the origin of emission and absorption spectra

Compare emission and absorption spectra with the continuous spectrum

Identify elements based on simple spectra

1 Distinguish between continuous, emission and absorption spectra.

_____

_____

_____

_____

2 Describe how emission spectra can be used to identify the presence of a substance.

_____

_____

_____

_____

3 When spectra are taken from distant stars (outside the Milky Way), it can be difficult to identify their composition because they are red shifted. Describe how this can be corrected.

_____

_____

_____

_____

4 Use the spectra below to answer the following questions.

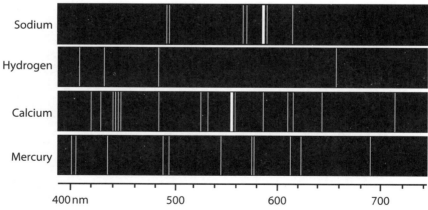

a An unknown sample was analysed and found to have two emissions around 490 nm. Suggest which of these elements could have produced the emission spectrum.

_____

b Identify two regions of calcium's spectrum that are unique.

_____

c A second unknown sample was found to have no emission in the 615 nm range. Identify the sample.

_____

5 Suggest why the spectra for heavier elements such as calcium and mercury are usually much more complex than for light elements such as hydrogen.

_____

_____

_____

6 Explain how stellar spectra gathered from devices in space are different from those gathered on Earth.

_____

_____

_____

_____

LEARNING GOALS

Identify features of stars

Classify stars according to their spectral features

Stars are classified according to their spectral features, including the presence of absorption lines, as well as their surface temperatures. The classification series starts with O, stars that have very high surface temperatures that give them a blue appearance, and progresses towards cooler stars through classifications B, A, F, G, K and M. These classifications are further broken down with numbers, roman numerals and letters. The numbers indicate how far along the scale the star sits. For example, a B1 star is hotter than a B2 star. The roman numerals represent a measure of the star's luminosity. The Sun, which is classified as G2V, can be interpreted as a 5500 K yellow star. It is one of the hotter stars in the G classification and median in luminosity.

1 Describe how a star generates an absorption spectrum despite being an emitter of radiation.

_____

_____

_____

2 Explain why the blue colour of a star indicates a higher surface temperature.

_____

_____

3 A star is found to have very broad but well-defined spectral lines. Explain what information this provides about the star.

_____

_____

_____

_____

4 Explain how greater density leads to greater spectral line broadening.

_____

_____

_____

_____

5 Describe the features identified by the following stellar classifications.

a O

_____

b M

_____

c K

_____

6   The table lists spectral features of the various classes of stars. Use the information in the table to classify the stars described in the following question parts.

Spectral type	Colour	Peak wavelength	Surface temperature (K)	Balmer line features	Other spectral features	Examples
O	blue	72	>30 000	weak	Ionised He^+ lines, strong UV continuum	Orionis C
B	light blue	145	11 000–30 000	medium	Neutral He lines	Achernar, Rigel, Spica
A	white	290	7500–11 000	strong	Strong H lines, ionised metal lines	Sirius, Vega
F	yellow-white	380	6000–7500	medium	Weak ionised Ca^+	Procyon, Canopus
G	yellow	530	5000–6000	medium	Ionised Ca^+, metal lines	Sun, Capella
K	orange	725	3500–5000	very weak	Ca^+, Fe, strong molecules, CH, CN	Arcturus, Aldebaran
M	red	960	<3500	very weak	Molecular lines, e.g. TiO, neutral metals	Betelgeuse, Antares
L	orange-brown	1200	<2700	(–)	Ionised He^+ lines, strong UV continuum	GD 165B
T	brown	4000	<900	(–)	Neutral He lines	Gleise 229B

a   6500 K surface temperature, yellow-white colour, presence of calcium absorption lines

_____

b   17 000 K surface temperature, blue colour, presence of neutral helium

_____

c   3000 K surface temperature, red, very weak Balmer series lines

_____

d   5500 K surface temperature, yellow colour, metal absorption lines present

_____

Describe the evolutionary stages of stars

Determine stellar features from Hertzsprung–Russell diagram points

Construct a Hertzsprung–Russell diagram from secondary data

1 Draw a flow chart to model the life cycle of a Sun-like star and a star of 10 solar masses.

2 Use the table from Worksheet 12.4 to identify two key spectroscopic features of:

a white dwarfs.

_____

b blue giants.

_____

c red giants.

_____

3 Classify the following stars.

a Red colour, very high luminosity

_____

b High temperature, high luminosity

_____

4 Explain how a cool red star can be far more luminous than a hot blue star.

_____

_____

_____

5 Plot the following stars on the HR diagram below and then use each star's position on the diagram to determine its spectral class and star group.

Star name	Absolute magnitude (M)	Temperature (K)	Spectral class	Star group type
Sun	4.8	5500		
Alpha Centauri A	4.3	5440		
Achemar	−2.4	20500		
Sirius A	1.4	9600		
Vega	0.5	9900		
Rigel	−7.2	12100		
Hadar	−5.3	25500		
Deneb	−7.2	9300		
Regulus	−0.8	13300		
Bellatrix	−4.3	23000		
Alpha Centauri B	5.8	4700		
Antares	−5.2	3300		
Pollux	1.0	4900		
Acrux	−4.0	28000		
Beta Centauri	−5.1	25500		
Polaris	−4.6	6100		
Sirius B	11.2	14800		
Procyon B	13.0	9700		
van Maanen's Star	14.2	13000		

H–R diagram

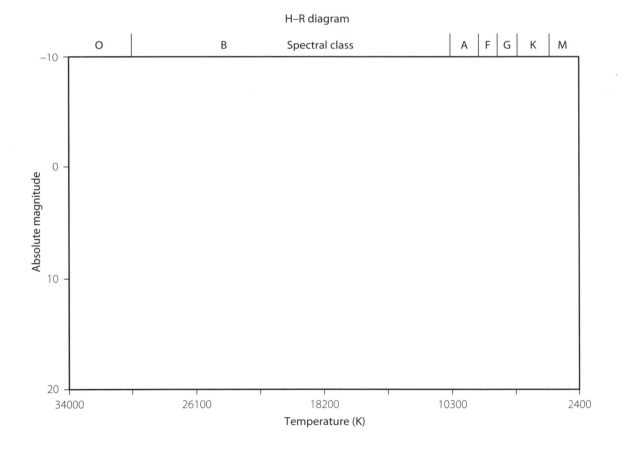

Relating the equivalence of energy and mass to the nuclear reactions that occur in stars

STUDENT BOOK
Pages 332–333

LEARNING GOALS

Calculate the mass defect for a given nuclear reaction

Calculate the binding energy for a given nuclear reaction

State numerical answers correct to an appropriate number of significant figures.

1   If the Sun loses 5.0×10^3 kg of mass each second, how much energy would this be equivalent to (assuming all mass lost is converted to energy through thermonuclear reactions)?

_____

2   An isolated neutron is unstable and will decay in a matter of minutes into a proton, an electron and an electron neutrino. Calculate the energy released in this decay process. (The mass of the electron neutrino is negligible in this calculation.)

_____

_____

3   Main sequence stars produce energy by turning four protons into a helium nucleus. They also produce two positrons in this process (mass 9.109×10^{-31} kg).

   a   If the mass of a helium nucleus is 6.646×10^{-27} kg, calculate the energy produced per nucleon produced.

   _____

   _____

   _____

   b   Calculate the energy produced per nucleon for the splitting of U-236 if the mass defect is 7.900×10^{-28} kg.

   _____

   _____

   _____

   c   Compare your answers in **a** and **b**. What can you conclude about the relative energy production of the fusion of hydrogen and the fission of uranium?

   _____

4   What two major forms of energy are released by nuclear reactions?

   _____

5   Using the concept of mass defect, explain the vast difference between the energy produced by chemical reactions and nuclear reactions.

   _____

   _____

   _____

LEARNING GOALS

Determine the conditions under which thermonuclear fusion occurs

Determine the process by which thermonuclear fusion occurs in the core of stars

Draw links between fusion process and the evolutionary stage of a star

The Sun produces almost all the energy that Earth and its inhabitants use. Every second the Sun converts over 5 tonnes of mass into approximately 4.5×10^{20} J of energy. This happens by thermonuclear fusion, the process of fusing light nuclei such as hydrogen into heavier nuclei.

The core of a star will undergo different nuclear reactions depending on the stage of its life cycle. Post-Main Sequence stars can undergo an enormous number of different reactions to produce elements up to and including iron (Fe).

Main Sequence stars will undergo the fusion of hydrogen to helium through one of two methods: the proton–proton (p–p) chain or the CNO cycle.

State numerical answers correct to an appropriate number of significant figures.

1 Describe the differing conditions under which the proton–proton chain and the CNO cycle occur.

_____

_____

_____

_____

_____

2 Explain why the CNO cycle requires higher temperatures and pressures than the p–p chain.

_____

_____

_____

3 Suggest why the two processes produce the same amount of energy.

_____

_____

4 Calculate the number of helium nuclei produced in the Sun's core each second to release 4.5×10^{20} J of energy. (You may want to use the answer from WS 12.6 question 3 to help solve this.)

_____

_____

_____

5 Compare the relative proportions of p–p chain and CNO cycle nuclear reactions present in the Sun, an A class star and an O class star, all of which are on the Main Sequence.

_____

_____

_____

_____

13 Structure of the atom

INQUIRY QUESTION: HOW IS IT KNOWN THAT ATOMS ARE MADE UP OF PROTONS, NEUTRONS AND ELECTRONS?

WS 13.1 **Investigating the evidence of the electron**

STUDENT BOOK
Pages 341–347

LEARNING GOALS

Identify experimental evidence for the properties of cathode rays

Outline Thomson's experiment to determine the charge/mass ratio of an electron

Outline how Millikan determined the fundamental charge

Today the existence of electrons is common scientific knowledge and is used to explain numerous phenomena, such as lightning, electric current, and radio transmitters and receivers. While evidence for the electron had mounted throughout scientific history, it was not until 1897 that JJ Thomson was able to demonstrate their presence. Although this was the first discovery of a fundamental particle, the electron was the second subatomic particle to be discovered (after the proton in 1886). Most experiments that provided evidence for the electron were from cathode ray tubes (CRT). The CRT is a glass tube with internal air pressure relatively close to zero. It also contains, at minimum, a cathode and an anode. Later models included additional parts to either investigate properties of, or make use of, cathode rays. To produce a cathode ray, a high voltage must be applied across the cathode and anode; this causes electrons to stream directly from the cathode to the anode with minimal deflection due to atmospheric particles.

State numerical answers correct to an appropriate number of significant figures.

1 Outline five properties of cathode rays as observed with the use of cathode ray tubes.

_____

_____

_____

_____

_____

_____

_____

_____

_____

_____

2 For two of the properties you listed in question **1**, assess the experiment leading to its discovery and explain why it led scientists to believe that cathode rays consisted of charged particles, specifically electrons.

_____

_____

_____

_____

_____

After Thomson's discovery of the electron in 1897, he investigated the properties of electrons using an evacuated tube with an inserted cathode and anode. To ensure the cathode rays travelled parallel to the tube he used a perforated anode (shaped like a doughnut) to absorb rays that were not perpendicular to the cathode surface. One of the well-known experiments he conducted was the charge-to-mass experiment. He hypothesised that electrons were small 'particles', and were affected by magnetic and electric fields, so he first designed an apparatus that would allow the magnetic and electric fields to be balanced against each other, which would be demonstrated by the electrons travelling in straight lines. That would require $F_E = F_B$, so he ensured the magnetic and electric fields acted perpendicularly, meaning that $F_B = qvB$. Given $F_E = qE$, we now have $qE = qvB$. Rearranging gives $v = \dfrac{E}{B}$, an expression for the velocity of the electrons.

Thomson then turned off the electric field and measured the radius of the arc the electrons were travelling in. Thinking of electrons as particles allows us to equate the magnetic and centripetal forces acting on the electron: $qvB = \dfrac{mv^2}{r}$. Rearranging gives us an expression for the charge-to-mass ratio: $\dfrac{q}{m} = \dfrac{v}{rB}$. Substituting in our previous expression for the velocity of the electron we get $\dfrac{q}{m} = \dfrac{E}{rB^2}$. The electric field strength was determined by the voltage and separation distance of the electric plates, and the magnetic field strength by the current and the number of coils in the electromagnets.

3 What assumption did Thomson make about electrons in order to derive the $\dfrac{q}{m}$ relationship?

_____

4 What could be concluded about the charge and/or mass of the electron from the $\dfrac{q}{m}$ value?

_____

_____

5 Draw a diagram to show Thomson's experimental set up. Label the key components.

6 The anode in Thomson's experiment was perforated to allow just a thin beam of electrons through. Suggest a reason for this.

_____

_____

7 Calculate the voltage required to produce an electric field strength of $80\,000\,\text{V}\,\text{m}^{-1}$ across a 5 mm plate gap.

_____

_____

8 If the effect of the electric field from question **7** on a cathode ray was cancelled out by a magnetic field with strength 0.45 T, calculate the speed at which the electrons were moving.

_____

_____

9 Using the charge-to-mass ratio of an electron, calculate the radius of curvature of an electron with velocity $2.1903 \times 10^{8}\,\text{m}\,\text{s}^{-1}$ when it is subject to a magnetic field of $7.419 \times 10^{-2}\,\text{T}$.

_____

_____

_____

Fourteen years after Thomson's discovery and determination of the charge-to-mass ratio of the electron, Robert Millikan determined the fundamental charge with the use of a simple and elegant experiment. In short, he induced a charge on tiny oil drops using X-rays, and used an electric field provided by oppositely charged parallel plates to balance the gravitational force acting on the oil drops and the electric force: $F_g = F_E$. He could calculate the mass of the oil droplet from its diameter and the electric field strength by the applied voltage and plate separation distance. This allowed him to develop an expression for the charge on each drop: $mg = \dfrac{Vq}{d}$. Rearranging gives $q = \dfrac{mgd}{V}$. Millikan observed that although the charges found on different drops were often different, they did not differ by random values continuously but rather differed by discrete multiples. He concluded that the amount of charge was quantised, and therefore there would be a fundamental charge equal in magnitude to the charge on an electron.

10 Describe the impact Millikan's results had on the understanding of the model of the atom at the time.

_____

_____

11 Explain how Millikan needed to set up the plates in his experiment to suspend the oil drops.

_____

_____

12 One oil drop was determined to be holding a charge of $-6.5 \times 10^{-19}\,\text{C}$. How could this be explained?

_____

_____

13 Assuming an oil drop had a net charge of $-1.6 \times 10^{-19}\,\text{C}$, calculate the mass of the droplet if it was suspended by an electric field of $65\,000\,\text{N}\,\text{C}^{-1}$.

_____

_____

14 Calculate the voltage required to suspend a 4.6275×10^{-16} kg oil droplet with a net charge of 1.92×10^{-18} C between two plates separated by 55 mm.

_____

_____

15 Outline a procedure for an investigation that could be used to model Millikan's experiment.

_____

_____

_____

_____

_____

_____

_____

_____

_____

16 Assess the use of models in physics education and understanding.

_____

_____

_____

_____

_____

_____

_____

_____

_____

LEARNING GOALS

Discuss the features and limitations of Thomson's and Rutherford's models

Outline the Geiger–Marsden experiment

In the late 1800s, Thomson developed the 'plum pudding' model of the atom, so named for its resemblance of a pudding (positive core) filled with plums (electrons) to make a relatively large and solid atom.

In 1911, under the guidance of Ernest Rutherford, Geiger and Marsden developed an experiment to confirm Thomson's model. They did this by firing alpha particles (helium nuclei) at an extremely thin layer of gold foil. Given that Thomson's atomic model implied a large radius and (therefore) low density, it was hypothesised that the small high energy alpha particles would pass through the foil with minimal to no deflection.

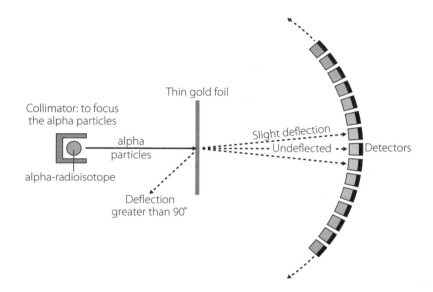

Although the vast majority of alpha particles were detected showing a small or no deflection, a small portion (approximately 1 in 8000) of the alpha particles were detected with deflection angles of 90° or more, some even at 180°. This led the group to conclude that the positively charged core of an atom must be extremely small and very dense. As a result, Rutherford developed his own model of the atom – a model that included a small positively charged nucleus that held most of the atom's mass, surrounded by electrons in a single fixed orbit. This left most of the atom as empty space.

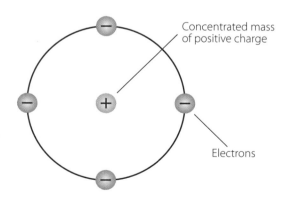

1 Identify a key difference between Rutherford's and Thomson's models.

_____

2 What was one of the most prominent limitations of Rutherford's model?

_____

_____

_____

3 Using Rutherford's model, explain the following observations from the Geiger–Marsden experiment.
   a No deflection of an alpha particle

   _____

   b A near 90° deflection

   _____

   c A near 180° deflection

   _____

4 What limitations did Thomson's model have in explaining the results of the Geiger–Marsden experiment?

_____

_____

5 What properties of alpha particles and gold nuclei did the Geiger–Marsden experiment utilise?

_____

_____

_____

The neutron proved to be an elusive particle to identify. It was not until 1932 that James Chadwick announced its discovery. He had been interested in the radiation produced by bombarding beryllium with high-energy alpha particles. This radiation was thought to be high-energy gamma photons, but that explanation could not explain the ability of the radiation to dislodge protons from paraffin wax, given the large mass of the proton. After 2 weeks of experimentation, Chadwick published his findings identifying this radiation as neutrons. He concluded, based on conservation of momentum, that these neutrons had a mass similar to that of the proton. This was the only way to conserve momentum in such a situation. He also concluded that neutrons had no net charge, allowing for much greater penetration into materials.

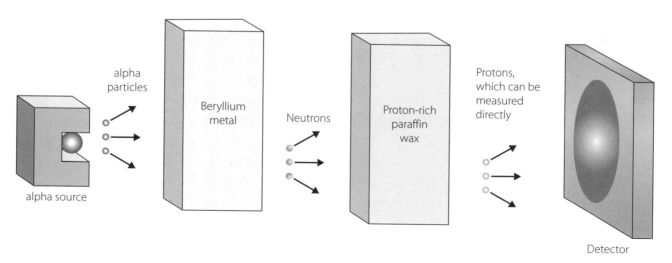

alpha source
alpha particles
Beryllium metal
Neutrons
Proton-rich paraffin wax
Protons, which can be measured directly
Detector

6 Write the nuclear equation for the bombardment of a beryllium nucleus by an alpha particle to produce carbon-12.

_____

7 Describe how Chadwick's discovery altered Rutherford's nuclear model of the atom.

_____

_____

_____

_____

8 Suggest why Chadwick believed that the radiation produced by bombarding beryllium with alpha particles could not be explained as high-energy gamma rays.

_____

_____

9 Outline some characteristics of free neutrons that may have made its discovery more difficult than the discovery of the electron.

_____

_____

10 Outline how the increasing evidence of subatomic particles has developed the nuclear model of the atom.

_____

_____

_____

_____

_____

_____

_____

Quantum mechanical nature of the atom

INQUIRY QUESTION: HOW IS IT KNOWN THAT CLASSICAL PHYSICS CANNOT EXPLAIN
THE PROPERTIES OF THE ATOM?

 ## Assessing the limitations of classical atomic models

STUDENT BOOK
Pages 358–360

LEARNING GOALS

Discuss limitations of Rutherford's model

Outline the contribution of Bohr

The Rutherford model of the atom made significant improvements over Thomson's model, namely the proposal of a small, dense, positively charged nucleus surrounded by empty space and electrons in an orbit with a large radius. Niels Bohr took a giant leap in the understanding of the structure of the atom by using the link between the wavelengths of hydrogen's spectral lines, suggested by Balmer, and the ideas linking electron energy and light energy put forward by Planck and Einstein.

1 Outline the limitations of Rutherford's model of the atom.

_____

_____

_____

2 State Bohr's postulates.

_____

_____

_____

_____

3 Describe how Bohr's model overcame some of the limitations of Rutherford's model.

_____

_____

_____

4 Explain how Bohr's contribution to the atomic model suggests that something other than classical physics was required to explain things at the very small scale.

_____

_____

_____

_____

5 Although Bohr's model was an improvement of previous models it still has its limitations.

   a Outline why his model could not be applied to atoms with more than one electron.

   _____

   _____

   _____

   _____

   b The Zeeman effect was another phenomenon that could not be explained by Bohr. Outline the Zeeman effect.

   _____

   _____

   _____

   c Identify two other spectral limitation of Bohr's model.

   _____

   _____

   _____

   _____

   _____

STUDENT BOOK
Pages 363–369

LEARNING GOALS

Recall that the energy of a photon depends on its frequency

Calculate the energy of a photon with a given wavelength or frequency

Use the Rydberg equation to determine wavelengths of specific transitions

Bohr used Planck's mathematical equation to demonstrate the quantum relationship between light and the atom, which suggested that the energy emitted and absorbed by an electron was proportional to the frequency of the photon emitted by the electron rather than to the intensity of the light.

The energy of a photon of light can be expressed as $E = hf$, where h represents Planck's constant. Using the inverse frequency–wavelength relationship, energy can also be expressed as $E = \dfrac{hc}{\lambda}$.

This equation is helpful in determining the energy of an electromagnetic wave with known frequency or wavelength, but it took the development of the Rydberg equation to determine the energy gap between Bohr's orbits. Although the Rydberg equation can be adapted for other elements, its simplest form can only be

used to predict the emissions from hydrogen. The equation is written as $\dfrac{1}{\lambda} = R_H \left(\dfrac{1}{n_f^{\,2}} - \dfrac{1}{n_i^{\,2}} \right)$ and has units of

wave number (m^{-1}). It relates the initial orbit (n_i) and the final orbit (n_f) and is scaled by the Rydberg constant (R_H). If the electron originates in a higher orbit, the equation will yield a positive result, indicating the emission of radiation. If the electron originates in a lower orbit, the equation will yield a negative result, indicating the electron absorbed radiation in order to make the transition.

State numerical answers correct to an appropriate number of significant figures.

1 Justify the statement that blue light has more energy than red light.

_____

_____

_____

2 Using $E = hf$, calculate the energy of each of the following emission lines of the hydrogen spectrum.

 a $H_\alpha = 656\,nm$

_____

_____

_____

 b $H_\beta = 486\,nm$

_____

_____

_____

c $H_\gamma = 434\,nm$

_____

_____

_____

d $H_\delta = 410\,nm$

_____

_____

_____

3 Why was the classical physics associated with Rutherford's model unable to explain the hydrogen absorption/
emission spectra?

_____

_____

4 Why was explaining spectra so important to the understanding of stars?

_____

_____

_____

_____

5 Predict the wavelengths of the first five transitions in the Lyman series.

> **HINT**
>
> The Lyman series is the transitions from n_i to $n = 1$.

_____

_____

_____

_____

_____

_____

_____

_____

_____

_____

_____

_____

6 a Electromagnetic radiation of 2.459×10^{15} Hz was found to be absorbed by a hydrogen atom. Given this absorption occurs in the UV part of the electromagnetic spectrum, show what transition this would correspond to.

b An electron in a hydrogen atom emits light of wavelength 97 nm. Assuming the electron ends in its ground state after it has emitted the radiation, show the initial state of the electron before the emission.

7 Provide two reasons why the Rydberg equation would need to be adjusted for other elements.

Explain the origin of simple spectra

Compare the features of different spectral series for hydrogen

Emission spectra are produced when electrons transition from a higher energy orbit to a lower energy orbit. During these transitions, the electron will emit a photon of light with energy equivalent to the energy gap between the initial and final orbits.

The orbits are not separated by equal amounts of energy, and the transition from $n = 2$ to $n = 1$ is the largest energy change. The gaps between subsequent levels become smaller and smaller as $n \rightarrow \infty$.

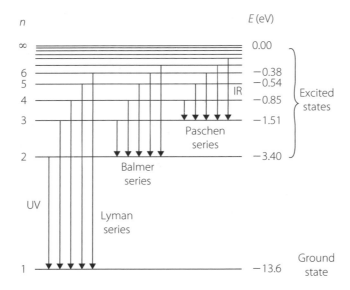

As shown in the diagram above, a number of different transitions contribute to an emission spectrum. Emissions from transitions to $n = 2$ are known as the Balmer series and are of particular interest because a portion of these emissions fall within the visible part of the spectrum and, thus, are detectable to the human eye with the use of a simple spectrometer.

Other common transitions such as those to $n = 1$ and $n = 3$ are known as the Lyman and Paschen series respectively. These emissions occur in the ultraviolet and infrared parts of the electromagnetic spectrum.

1 To determine either emission or absorption spectral lines, a spectroscope must be used.

a Describe why the human eye cannot detect spectral lines.

_____

_____

_____

b Describe how a spectroscope allows specific spectral lines to be observed.

_____

_____

_____

2 Explain why only four lines are observed in a spectroscope when viewing hydrogen's emission spectrum.

_____

_____

3 Using the diagram at the start of this worksheet, what conclusions can be drawn about the relative energy gaps between orbits from the frequencies produced in the hydrogen spectrum?

_____

_____

_____

_____

4 Explain why the transition between two energy levels results in a photon of the same energy being emitted as absorbed.

_____

_____

_____

_____

5 Melika and Franco were conducting an experiment to determine the wavelength of light produced by hydrogen as an emission spectrum. The results suggested that the H_α line (with an accepted wavelength of 656 nm) was equal to 672 nm.

a Identify a risk and an associated precaution to be taken when carrying out spectroscopic observations.

_____

_____

_____

b Describe a source of random and a source of systematic error that could lead to this discrepancy.

_____

_____

_____

WS 14.4 Investigating the evidence of matter waves

STUDENT BOOK
Pages 369–378

In Einstein's explanation of the photoelectric effect, he assumed that light waves acted as a particle (photon) in order to explain the quantisation of energy. Louis de Broglie used Einstein's energy equations to derive an expression relating momentum and wavelength. By combining $E = mc^2$ and $E = \dfrac{hc}{\lambda}$, wavelength can be represented as $\lambda = \dfrac{h}{mc}$. de Broglie further hypothesised that, by substituting the velocity of an object such as an electron or neutron for the speed of light, this relationship could be used to determine the de Broglie wavelength of any moving object.

After de Broglie hypothesised the existence of matter waves in 1924, numerous pieces of experimental evidence have been found to support this theory, such as the diffraction patterns produced by Davidson and Germer and by GP Thomson. The diffraction patterns were produced by scattering a beam of electrons through the crystalline lattice of metals. These diffraction patterns, as well as those produced by the scattering neutrons, are still used today in determining crystal structures.

The implications of de Broglie's theory were that both particle and wave nature could be exhibited and observed from the same entity, whether it be a photon of light or an electron. This changed the understanding of electrons from being planet-like in their orbit of the nucleus to existing as a standing wave around the nucleus. This was a step in further explaining the stability of electron orbits as well as the forbidden energies between stable orbits.

State numerical answers correct to an appropriate number of significant figures.

1 Although it is possible to calculate the de Broglie wavelength of any object providing its velocity is not zero, it is not possible to measure the wavelength of objects such as humans or cars experimentally. Explain why.

_____

_____

_____

2 Describe how de Broglie's theory supported Bohr's stable orbit theory.

_____

_____

_____

_____

3 Outline two pieces of evidence that supported de Broglie's theory.

_____

_____

_____

_____

4 Calculate the wavelengths of the following particles moving at a speed of $2.52 \times 10^6 \, \text{m s}^{-1}$.

a An electron

b A neutron

c A helium nucleus (mass $= 6.644\,77 \times 10^{-27} \, \text{kg}$)

5 The resolution of an image is dependent on the wavelength of the incident waves: the shorter the wavelength, the higher the resolution. Explain quantitatively why an electron microscope would have a higher resolution than a light microscope.

Analysing the contribution of Schrödinger to the current atomic model

STUDENT BOOK
Pages 378–380

LEARNING GOALS

Analyse the contribution of Schrödinger to the atomic model

Classical physics was becoming more obsolete in the explanation of subatomic particles. The quantisation of energy put forward by Einstein, Planck and Bohr was further developed by Pauli, Heisenberg (not covered in the HSC course) and Schrödinger.

Schrödinger developed an equation from which solutions could be derived that would determine the probability of finding an electron in a particular position. These solutions showed that electrons did not sit in discrete individual orbits but instead could exist anywhere within a specified region, known as an orbital. Orbitals existed with different energies and shapes, and electrons could transition between specific orbitals by absorbing or emitting radiation. However, the precise location of the electron within an orbital could not be known.

1 Describe an orbital.

_____

_____

2 How do Schrödinger's orbitals differ from Bohr's orbits?

_____

_____

_____

3 Summarise the contribution of Schrödinger to the atomic model.

_____

_____

_____

_____

4 Draw a diagram of a quantum model of the atom.

5 While Schrödinger supported a model in which electrons could be thought of as both particles and waves, he understood that this model could not be applied to the macroscopic world. As a result, he developed a thought experiment known as 'Schrödinger's cat'.

a Outline the 'Schrödinger's cat' thought experiment.

_____

_____

_____

_____

b Explain how the thought experiment links to the quantum mechanical atomic model.

_____

_____

_____

c Using the Schrödinger's cat thought experiment, explain why it is inappropriate to apply quantum mechanical thought processes to the macroscopic world.

_____

_____

_____

INQUIRY QUESTION: HOW CAN THE ENERGY OF THE ATOMIC NUCLEUS BE HARNESSED?

WS 15.1 Analysing the spontaneous decay of unstable nuclei

STUDENT BOOK
Pages 384–391

LEARNING GOALS

Determine the cause of instability of a nucleus

Describe the processes by which nuclei improve their stability

Identify the products of common decay methods

The stability of atomic nuclei depends on a large number of factors, including, but not limited to, the overall size and mass of the nucleus, their proton-to-neutron ratio and their level of excitation.

Nuclei with $Z > 82$

All nuclei with more than 82 protons are unstable due to their large number of protons (more specifically, the particles inside these protons). When the coulombic repulsion force between these particles becomes greater than the strong nuclear force holding the nucleus together, the nucleus becomes unstable and subject to alpha decay (the emission of a helium nucleus from the parent nucleus).

Proton-to-neutron ratio

Although alpha decay is almost exclusive to nuclei heavier than lead, beta decay occurs as a result of the proton-to-neutron ratio being too high or too low. A ratio that is too low (also known as proton deficiency) results in the decay of a neutron into a proton and an electron via a process known as β^- decay. A ratio that is too high (neutron deficiency) results in the decay of a proton to a neutron with the emission of a positron via β^+ decay.

Excited states

Nuclei, like electrons, can exist in a ground state or in a number of excited states. Also, like electrons, when nuclei relax back to their ground state they emit a photon of electromagnetic radiation. Given the size of a nucleus compared to an electron, the radiation produced is within the gamma part of the spectrum rather than the visible. When a nucleus undergoes transmutation by alpha or beta decay, there is often also an emission of gamma radiation as the daughter nucleus relaxes back to its ground state.

The line of stability

Any nucleus that does not fit into one of the three above categories is considered to be on the line of stability. The gradient of this line provides the ratio of neutrons to protons required for stability but ends at $Z = 82$. For the first 20 elements the gradient is approximately 1. As the Z number increases beyond 20, the number of neutrons required per proton to mediate the electrostatic forces between protons increases to approximately 1.5 neutrons for each proton.

1   Summarise the circumstances under which each of the decay methods (i.e. alpha, β^+ and β^-, and gamma) are likely to occur.

_____

_____

_____

_____

_____

2 Heavy isotopes such as U-238 often undergo beta and gamma decay in addition to alpha decay as they decay towards stability. Explain the cause of these decay processes.

_____

_____

_____

_____

_____

_____

3 Write the balanced nuclear equation for the alpha decay of thorium-229.

_____

4 Write the balanced nuclear equation of the beta minus decay of carbon-14.

_____

5 In recent times, nuclei as large as Oganesson ($Z = 118$) have been discovered/developed. Suggest why the discovery/development of new elements is becoming more difficult.

_____

_____

_____

6 Provide an equation for the decay of the following isotopes.

   a Beta negative decay of $^{99}_{43}\text{Tc}$

_____

   b Beta positive decay of $^{23}_{12}\text{Mg}$

_____

   c Alpha decay of $^{219}_{86}\text{Rn}$

_____

7 Complete the following equations.

   a $^{228}_{90}\text{Th} \rightarrow$ _____ $+\ ^{4}_{2}\text{He}$

   b _____ $\rightarrow\ ^{131}_{54}\text{Xe} +\ ^{0}_{-1}\text{e}$

   c $^{11}_{6}\text{C} \rightarrow\ ^{11}_{5}\text{B} +$ _____

LEARNING GOALS

Define half-life

Outline the concept of a decay constant

Calculate the half-life or decay constant from data provided

The decay of a nucleus is a random process that cannot be predicted. If a single nucleus of an unstable isotope is isolated, there is no way of determining when it will decay. However, for a statistically significant number of nuclei, a half-life can be determined. The half-life of any given isotope is defined as the time after which one half of a sample would have decayed.

The half-life, referred to as $t_{1/2}$, can be used to calculate λ, the decay constant of the isotope through the equation $\lambda = \dfrac{\ln 2}{t_{1/2}}$. The decay constant is unique to each radioisotope and is the probability that a radionucleide will decay per unit time. This constant can be used to calculate the number of undecayed nuclei at any time after t_0 by using the equation $N_t = N_0 e^{-\lambda t}$. In this equation, N_t is the number of undecayed nuclei at time t, N_0 is the original number of undecayed nuclei, λ is the decay constant as defined above, and t is time that has elapsed.

State numerical answers correct to an appropriate number of significant figures.

1 Technetium-99 m has a half-life of 6.0058 hours. Determine what mass of technetium-99 m would be left after 36.0348 hours from a 345 g sample.

_____

_____

_____

2 Cobalt-60 has a half-life of 5.27 years. Calculate its decay constant.

_____

_____

_____

HINT

Note: one year = 365.25 days.

3 Rn-222 has a decay constant of $2.1\,\mu s^{-1}$.

a Calculate its half-life.

_____

_____

_____

b Compare both qualitatively and quantitatively the activity of Co-60 and Rn-222.

_____

_____

_____

4 Polonium-210 has a decay constant of $5.81 \times 10^{-8}\,s^{-1}$. How many nuclei would be left after 2.65 years if the sample originally contained 1 mole (6.022×10^{23} atoms) of polonium?

_____

_____

5 An unknown isotope X was found to have 72.045% of its nuclei decay in 46.21 hours. Calculate the decay constant of X.

6 A sample of a suspected alpha emitter was placed inside an alpha detector. Over the space of 15.00 years, 1375 alpha emissions were detected. Assuming the daughter nuclei of this decay are stable and given there were $1.500\,000\,000\,0 \times 10^8$ nuclei at the start of the experiment, calculate the predicted half-life of this isotope.

Explain the process of nuclear fission

Compare controlled and uncontrolled nuclear reactions

Most radioisotopes achieve stability through a single or series of α, β and γ decays. Some are so unstable that the nucleus splits into two or more smaller daughter nuclei in a process known as nuclear fission. Fission is only possible in nuclei with a Z number greater than 26 (iron) and is more common in isotopes with $Z > 90$. The most common fissile isotope is U-235.

Modelling fission

Sustained nuclear fission was first achieved by Enrico Fermi in 1942 in his Chicago pile-1, a very primitive nuclear reactor. As nuclear fission cannot be recreated in the school laboratory, modelling becomes an effective way to understand some of the key features. One such model is known as the mouse trap model. This model uses ping-pong balls to represent neutrons and mouse traps to represent the fissionable nuclei, and involves placing and setting hundreds of spring-loaded mechanical mouse traps, each with a ping-pong ball on top. One of the mouse traps is set off, which launches a ping-pong ball; when it lands on another trap it triggers that trap as well. This process continues until all traps had been triggered. In some instances a ping-pong ball sets off more than one additional trap, increasing the rate at which traps are being triggered.

Mechanisms of fission

Nuclear fission occurs when the electrostatic repulsion between protons within a nucleus overcomes the strong nuclear force of attraction between them. Fission can occur spontaneously in rare cases, but it is usually initiated by neutron bombardment. In the case of U-235, a neutron is captured, producing highly unstable U-236. This then splits into two daughter isotopes, also producing additional neutrons. The fission of U-235 can occur by a number of pathways, one of which is shown below.

$$^{235}_{92}\text{U} + ^{1}_{0}\text{n} \rightarrow ^{144}_{56}\text{Ba} + ^{89}_{36}\text{Kr} + 3\,^{1}_{0}\text{n}$$

In this example, each of the three neutrons produced has the potential to be captured by nearby nuclei to initiate another split. Under the right circumstances, this creates a chain reaction, leading to an uncontrolled release of energy. When a critical mass of a fissile material is present, it creates a uncontrolled chain reaction, leading to an explosion, as used in nuclear warheads.

The rate of this chain reaction can be controlled with the use of control rods (made of a neutron-absorbing material such as cadmium). These control rods can be inserted to separate segments of the mass of fissile isotope to reduce the rate of reaction and control its energy output. In this scenario, fission can be used in a nuclear reactor to generate power.

Energy production

The energy produced in a fission reaction takes several forms. Most of the energy is carried away in the kinetic energy of the daughter nuclei. The remaining energy is carried away through gamma rays and the kinetic energy of neutrons and other subatomic particles. In total, the average energy output of the reaction is approximately 200 MeV. This energy comes about as a result of the difference between the rest mass of the parent nuclei and the products of the reaction.

1 Explain what is meant by an uncontrolled nuclear reaction. Refer to a specific reaction in your answer.

_____

_____

_____

_____

_____

2 How many neutrons per reaction would be required to be absorbed by subsequent nuclei to maintain a controlled reaction? Explain how this is achieved.

_____

_____

3 What form does most of the energy take in a fission reaction?

_____

4 Given that not all product neutrons are captured by neighbouring nuclei in nuclear reactions, suggest what would need to occur in a nuclear fission reaction in order for a controlled chain reaction to be sustained.

_____

_____

_____

_____

5 What did the ping-pong balls represent in the mouse trap fission model?

_____

6 Discuss the benefits and limitations of this model.

_____

_____

_____

_____

_____

_____

_____

WS 15.4 Analysing the conservation of mass–energy in nuclear transmutations

Describe the concept of binding energy

Identify the conditions under which fusion occurs

To say that a nuclear reaction produces energy is not an accurate account of what is happening. In every nuclear transmutation, whether it be radioactive decay, fission or fusion, the energy 'produced' is actually the result of a transformation of energy from mass to other forms of energy according to Einstein's equation $E = mc^2$. In fact, this is true for all chemical reactions as well. Given the magnitude of the speed of light, the amount of mass that is converted to energy is exceedingly small and is often considered negligible in chemical reactions.

All nuclear transmutations must conserve mass–energy. The transmutations that occur without external input are considered to be spontaneous. Examples include radioactive decay (including alpha, beta and gamma) as well as fission of nuclei with a very high Z number, such as Cf-252. Transmutations that require bombardment by other particles are considered to be artificial transmutations and include the fission of U-235 (neutron bombardment) and the fusion of hydrogen (proton bombardment).

1   Define spontaneous and artificial transmutation and provide an example of each, including an equation.

_____

_____

_____

_____

2   Why do nuclear power stations use only isotopes that undergo artificial transmutation as a fuel?

_____

_____

_____

_____

Stars undertake thermonuclear fusion to power their existence. To complete this process, they must create temperatures in excess of 10 million kelvin, and pressures in the order of 250 billion atmospheres. These conditions allow light nuclei such as protons to overcome their electrostatic repulsion and fuse to form a new nucleus in a process governed by the strong nuclear force. When two nucleons fuse in this way, the result is more stable (and therefore lower in energy) than a single nucleon on its own. This energy difference, known as the binding energy, is what makes fusion such a powerful tool for converting mass to energy. The binding energy per nucleon is defined as zero for a single nucleon and (generally) increases as nucleons are added, up to the iron isotopes, which have the highest binding energy per nucleon (i.e. are the most stable). This is why even the biggest stars will not make any elements heavier than iron during their fusing lifetime.

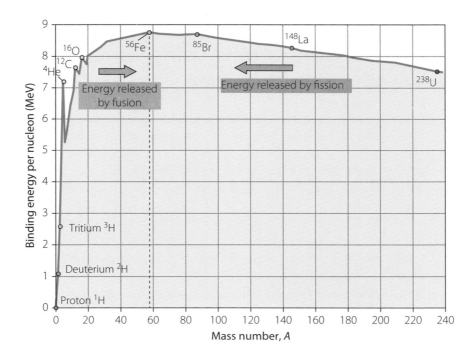

The greater the difference between the binding energy per nucleon of the products and reactants, the greater the energy produced by their fusion. The binding energy of any nucleus can be determined by calculating the rest mass of the nucleus and subtracting it from the rest mass of its individual constituents. This provides us with the mass defect, which can be substituted into Einstein's mass–energy equivalence equation.

3   Give two reasons why nucleons are required to be subjected to such high temperatures and pressures for fusion to take place.

_____

_____

_____

4   Suggest why stars do not produce elements such as lead as part of their normal fusion cycle.

_____

_____

_____

5   Explain what would be required in order to fuse nuclei to form elements heavier than iron. Provide an example of where these reactions could be found.

_____

_____

_____

6   Why does a single proton have zero binding energy?

_____

7   Using the graph above, what can be concluded about the relative amount of energy released by the fusion of light nuclei compared to the fission of heavy nuclei?

_____

_____

_____

Calculate the energy released during nuclear reactions in both joules and electron volts

When using the conservation of mass–energy to predict the energy release from nuclear transmutation, it is often more convenient to use units for mass and energy that are much smaller than the standard kilogram and joule.

The mass of atomic nuclei and nucleons is often quoted in atomic mass units (u). One u is $\dfrac{1}{12}$ the mass of a carbon-12 nucleus.

The electron volt is a unit that we first came across in the Year 11 course. It is defined as the energy gained by an electron as it moves through a potential difference of 1 V.

While the kilogram and joule remain the SI units for mass and energy, the mass defect and binding energy of a nucleus are often calculated and quoted in u or eV. One u is equal to $\dfrac{931.5\,\text{MeV}}{c^2}$. All of these values can be found on the HSC Physics Data Sheet.

Use the following tables to answer the questions below.

Particle	Rest mass (kg)
Proton	1.673×10^{-27}
Neutron	1.675×10^{-27}
u	1.661×10^{-27}
Electron	9.109×10^{-31}

Nucleus	Rest mass (u)	Nucleus	Rest mass (u)
Ba-144	143.923	Th-230	230.033
He-4	4.0026	Ra-226	226.025
Kr-89	88.917	Rn-222	222.018
C-14	14.003	Po-218	218.009
Si-28	27.977	Po-214	213.995
Fe-56	55.935	Po-210	209.982
U-238	238.051	Bi-214	213.999
U-235	235.044	Bi-210	209.983
U-234	234.041	Pb-214	214.000
Pa-234	234.043	Pb-210	209.984
Th-234	234.044	Pb-206	205.973

State numerical answers correct to an appropriate number of significant figures.

1 Calculate the mass defect in the reaction $^{235}_{92}\text{U} + ^{1}_{0}\text{n} \rightarrow ^{144}_{56}\text{Ba} + ^{89}_{36}\text{Kr} + 3\,^{1}_{0}\text{n}$.

_____

_____

_____

2 Calculate the binding energy of a helium nucleus in joules, given its rest mass is $6.644\,77 \times 10^{-27}\,\text{kg}$.

_____

_____

_____

3 If the Sun converts $5.00 \times 10^{9}\,\text{kg}$ of mass each second, calculate the energy released in the core in 24 hours.

_____

4 Calculate the binding energy of silicon-28. Give your answer in eV.

_____

_____

_____

5 Calculate the binding energy per nucleon of iron-56. Give your answer in eV.

_____

_____

_____

6 Uranium-238 undergoes a series of decay steps to achieve stability at lead-206. One series of decay is shown below.

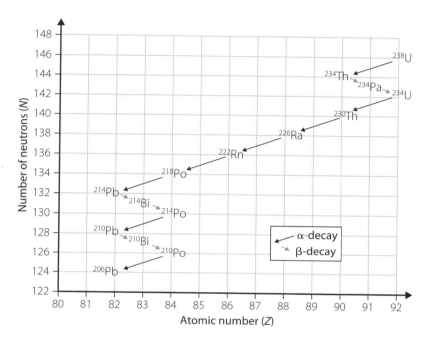

a Calculate the energy in joules released during two alpha decay steps and two beta decay steps.

_____

_____

_____

_____

_____

_____

_____

_____

_____

_____

_____

_____

_____

_____

_____

_____

_____

_____

_____

_____

_____

_____

_____

_____

_____

b Which decay steps release the most energy? Suggest a reason why.

_____

_____

_____

9780170449687

INQUIRY QUESTION: HOW IS IT KNOWN THAT HUMAN UNDERSTANDING OF MATTER IS STILL INCOMPLETE?

WS 16.1 Analysing the evidence of further subatomic particles

STUDENT BOOK
Pages 413–418

LEARNING GOALS

Describe the evidence for the existence of subatomic particles other than electrons, protons and neutrons

Outline theoretical and experimental evidence for antiparticles

Scientific models are developed as a result of experimental evidence. Just as quickly as evidence for the existence of the proton and neutron surfaced, there was mounting evidence that these could be broken down into more simple components and were, therefore, not fundamental like the electron.

Antimatter

In the 1920s, Paul Dirac, produced two solutions to Schrödinger's equations that described not only the electron but also an identical particle with opposite charge. It was theorised that when these particles interacted, they would annihilate each other, converting their combined mass to energy according to Einstein's mass–energy equivalence equation. This particle (positron) was first observed by Carl Anderson in 1932.

1   What does the discovery of the anti-electron suggest about other subatomic particles?

_____

_____

2   Given that the annihilation of an electron–positron pair produces two identical gamma rays, calculate the wavelength of these waves.

_____

_____

_____

_____

Beta decay

Unstable nuclei undergo spontaneous transmutation via alpha, beta or gamma decay. Beta decay is more complex than alpha or gamma as it can occur as β^- or β^+. β^- decay involves the emission of an electron from the nucleus, and β^+ is the emission of a positron.

3   Suggest what information could be used to hypothesise that the neutron was not a fundamental particle.

_____

_____

_____

4 Explain what changes occur within a nucleus when it undergoes β⁻ decay. Include a nuclear equation.

_____

_____

_____

5 Explain what changes occur within a nucleus when it undergoes β⁺ decay. Include a nuclear equation.

_____

_____

_____

6 What does beta decay suggest about protons and neutrons?

_____

_____

7 Free neutrons are unstable, decaying into a proton and electron with a half-life of about 15 minutes. Free protons, on the other hand, are extremely stable and do not decay into a neutron and positron. Suggest a reason why this does not occur.

_____

_____

_____

Cosmic rays

Cosmic rays are high energy atomic nuclei (90% protons, 9% alpha particles and 1% heavier nuclei) that emanate from the Sun. When they hit Earth's atmosphere, they produce a number of decay products, called secondary cosmic rays, that are easily detected at Earth's surface. These particles include muons and mesons, particles that were not accounted for in the Rutherford–Bohr atomic model.

Neutrinos

In 1959 Clyde Cowan and Fred Reines discovered the neutrino, a particle that had previously been theorised, with no electrical properties and near-zero mass. These particles were incredibly small and very weakly interacting with matter, making their detection difficult.

8 Summarise, using specific examples, the evidence suggesting that protons and neutrons were not fundamental particles and that other subatomic particles exist.

_____

_____

_____

_____

_____

_____

_____

_____

_____

_____

9 Describe the impact the discovery of these other subatomic particle has had on the understanding of matter.

_____

_____

_____

STUDENT BOOK
Pages 421–425

Identify the fermions and bosons that make up the Standard Model of matter

Classify particles as fermions or bosons

Classify groups of particles as hadrons, baryons or mesons

Distinguish between fundamental particles and subatomic particles

The evidence put forward by numerous physicists indicated that a far bigger model than just 'the atom' was needed to explain the subatomic world. Now known as the 'Standard Model of matter', this model is used to explain (or partially explain) all fundamental particles and the forces acting on them. A graphical representation of this model is shown below.

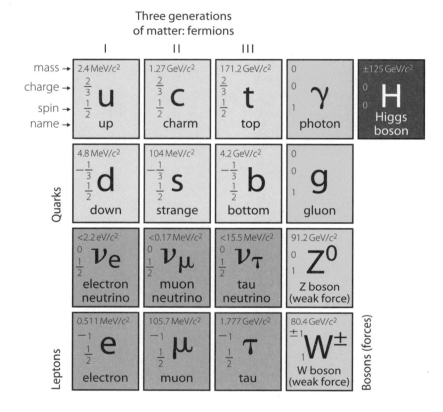

The model is split into fermions and bosons. Fermions have half integer spin $\left(\pm\frac{1}{2}\right)$ and each fermion has a

corresponding antiparticle. All fermions obey the Pauli exclusion principle. Bosons have integer spin (-1, 0 or $+1$) and are responsible for mediating forces between particles.

Quarks

Coming in six flavours, quarks are the building blocks of most matter. Regular matter (such as protons and neutrons) is comprised of up and down quarks. The second- and third-generation quarks are generally only found in higher energy situations, such as the Big Bang and the environments created within particle accelerators.

Hadrons

While leptons do not combine with other particles, quarks come together in pairs or triplets to form other particles collectively called hadrons. A triplet of quarks forms a baryon. Three up and down quarks (either two ups and one down or one up and two downs) form protons and neutrons respectively.

Mesons are produced when a quark and anti-quark pair up. The pi mesons (or pions) are produced when the pair consists of only up and down quark/antiquarks.

Many other baryons and mesons exist, formed by adding second and third generation flavours, but these are present only in high energy situations.

Leptons

The leptons are another group of fundamental particles and include the electron. Like the up and down quarks, the electron is present in regular matter. Muons are higher energy particles typically produced when cosmic rays interact with the upper atmosphere. The tau is the most massive of the leptons and is highly unstable. It is typically only found in the decay of subatomic particles. There are only three flavours of leptons, but each flavour is accompanied by a neutrino when it is produced.

In addition to the 12 matter particles mentioned above, every flavour has its own antiparticle, meaning 24 fundamental particles make up all matter in the Universe, according to the current model.

Bosons

In addition to the 24 fundamental particles, there are four fundamental forces within the Standard Model of matter. Each of these forces is believed to be mediated by a particle known as a boson. The strong nuclear force is mediated by the gluon. The weak nuclear force is mediated by the Z and W bosons, and the electromagnetic force is mediated by the photon. Note that there is no gravity-mediating particle in the model. At the current time this particle (graviton) is theorised but has not yet been confirmed.

The most recently discovered boson is the Higgs boson, believed to mediate all matter with the Higgs field. It is this mediation that gives all fundamental particles their rest mass.

1 Define boson.

_____

_____

_____

2 Describe the role of the strong nuclear force.

_____

_____

_____

3 Describe the difference between a fundamental particle and a subatomic particle.

_____

_____

_____

4 Describe the difference between a baryon and a meson.

_____

_____

_____

_____

5 Provide an example of a hadron and identify its components.

_____

_____

6 Explain why it is so difficult to observe second- and third-generation quarks.

_____

_____

_____

7 What would be the fundamental particles of an antimatter hydrogen atom?

_____

_____

_____

Investigating the importance of particle accelerators in obtaining evidence to further scientific theories

STUDENT BOOK
Pages 419–421

Describe the structure and operation of particle accelerators

Explain the role of particle accelerators in particle physics

Many of the discoveries that have contributed to the development of the Standard Model of matter have been made using particle accelerators. These allow scientist to create high energy environments and detect the decay products of these interactions.

Particle accelerators are generally built in one of three designs, called a cyclotron, synchrotron or linear particle accelerator.

Cyclotrons

The cyclotron uses a large pair of disc magnets to create a magnetic field and two hollow D-shaped structures to create an electric field. Since there is no electric field inside any hollow charged object, the electric field only exists in the gap between the Ds. A charged particle such as a proton is injected near the centre of the cyclotron. The electric field accelerates the particle across the gap while the magnetic field exerts a force on the particle resulting in it undergoing uniform circular motion until it reaches the region of electric field again. At this point the polariy of the Ds switches and the particle accelerates across the gap again. Each acceleration results in an increase in velocity which causes the radius of the circle traced by the particle in the magnetic field to increase. Once sufficient speed is reached and the radius is sufficiently large it will exit the accelerator and strike a target.

Cyclotrons are more compact than linear particle accelerators and this makes them more versatile in their applications, which include cancer treatment. However, their ability to accelerate charged particles to high speeds is limited and so they are restricted in their ability to provide information of the highest energy interactions.

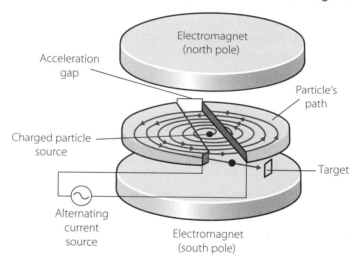

Linear particle accelerators

As their name suggests, these accelerate particles in a straight line. They employ alternating electric fields between successive cylinders. To compensate for the increase in speed, the cylinders (drift tubes) increase in length the further along the accelerator they are.

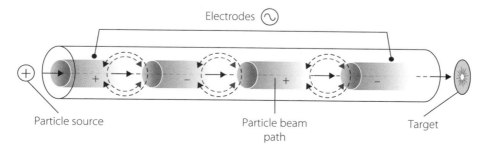

Linear accelerators are much simpler in design than a cyclotron, but they need significant space to accelerate particles to high energies. They are also limited in their ability to provide information about the highest energy interactions.

Synchrotrons

Synchrotrons take the best properties of the previous two designs to achieve the highest energy collisions. They use the simple accelerating principles of a linear particle accelerator but use a doughnut arrangement of the tubes that allows the particles to experience greater time periods under the accelerating forces of electric fields. The particles complete multiple circuits of the accelerator to achieve extremely high energies and thus delve deeper into the subatomic world than ever before. The circular motion is produced by multiple magnetic fields throughout the accelerator. The Large Hadron Collider (LHC) at CERN is an example of a synchrotron.

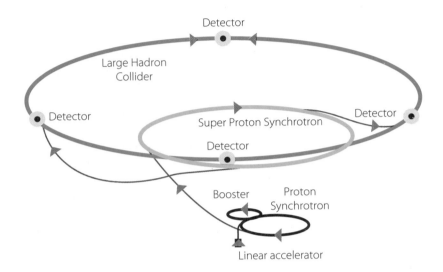

Synchrotrons typically use multiple loops to allow not only the acceleration of charged particles but also the storage of these particles at high energies so that collisions can take place when a sufficient number of particles exist or at a time suited to the detection equipment.

State numerical answers correct to an appropriate number of significant figures.

1 Why could a neutron not be used in a particle accelerator?

_____

_____

_____

2 Explain why synchrotrons achieve much higher energies than cyclotrons or linear particle accelerators.

_____

_____

_____

3 Calculate the relativistic momentum of a proton that had been accelerated to 0.9999999c in the LHC.

_____

_____

_____

4 Suggest reasons why particle accelerators need to be evacuated.

_____

_____

_____

_____

5 Explain why the particles cannot be accelerated to the speed of light.

_____

_____

_____

6 The diagram of the linear particle accelerator above shows the drift tubes. Explain why these tubes must increase in length at larger distances from the ion source.

_____

_____

_____

7 What potential difference between charged plates, placed 14 mm apart, would be required to exert a force of $7.1 \times 10^{-14}\,\text{N}$ on a proton?

_____

_____

_____

Module eight: Checking understanding

Circle the correct answer for questions **1–5**.

1 In one second the Sun generates 3.8×10^{26} joules. About how much mass does the Sun lose to do this?

    A   1 billion tonnes per second

    B   700 million tonnes per second

    C   6.81 million tonnes per second

    D   4.22 million tonnes per second

2 What essential feature of the structure of matter allows us to use the atomic spectra of elements to identify them?

    A   Electron orbits have quantised energy values.

    B   The number of electrons is different for each element.

    C   The number of electrons that can fit into any orbit is given by $2n^2$.

    D   Electrons can absorb or release energy when they change levels.

3 Millikan's oil drop experiment was important because:

    A   it quantified the charge on an electron.

    B   it established the charge-to-mass ratio of the electron.

    C   it used a simple apparatus to equate gravity and electrostatic charge.

    D   it enabled the calculation of the mass of a proton when combined with previous data from JJ Thomson.

4 C-14 is an unstable isotope of carbon. What is the most likely method by which it will decay?

    A   Alpha

    B   Beta positive

    C   Beta negative

    D   Gamma

5 Which of the following is NOT a group on the Hertzsprung–Russell diagram?

    A   Main sequence

    B   Blue dwarf

    C   Red dwarf

    D   Blue giant

State numerical answers correct to an appropriate number of significant figures.

6 Outline the causes of the instability that lead to alpha, beta and gamma decay.

_____

_____

_____

_____

7 Explain the impact of cepheid variables on the determination of the expanding Universe.

_____

_____

_____

_____

8 Explain how stellar spectra can be used to identify the chemical composition of stars.

_____

_____

_____

_____

_____

9 Compare the proton–proton chain and CNO cycle.

_____

_____

_____

_____

_____

10 Outline the properties of cathode rays.

_____

_____

_____

11 Describe the implications of the Geiger–Marsden experiment.

_____

_____

_____

_____

12 State Bohr's postulates.

_____

_____

_____

_____

13 Derive de Broglie's relationship between wavelength and momentum (not required knowledge according to syllabus) and identify one piece of evidence that supports this theory.

_____

_____

_____

_____

14 Describe how Schrödinger's ideas changed Bohr's model of the atom.

_____

_____

_____

15 Compare cyclotrons and synchrotrons.

_____

_____

_____

_____

_____

_____

16 Distinguish between a fermion and boson and provide an example of each.

_____

_____

17 Write an equation for the alpha decay of U-238.

_____

18 Calculate the wavelength of light produced by an electron transition from $n = 6$ to $n = 4$.

_____

_____

_____

19 Calculate the decay constant of Tc-99 given its half-life of 6 hours.

_____

_____

_____

20 Calculate the radius of curvature of an electron travelling at $6.25 \times 10^6 \, \text{m s}^{-1}$ when placed in a magnetic field of strength $1.972 \times 10^{-4} \, \text{T}$.

_____

_____

_____

Section 1

20 marks

Circle the correct answer for questions **1–20**.

1 The diagram below shows the path of a projectile in an environment with zero air resistance.

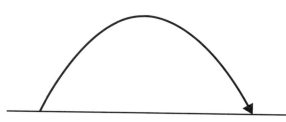

Which series of vector arrows best illustrates the forces acting on the projectile at various points during the flight?

A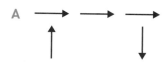

B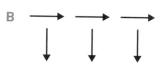

C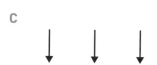

D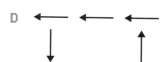

2 The diagram below shows a representation of a wave. Which of the following is the best identification of the concept shown?

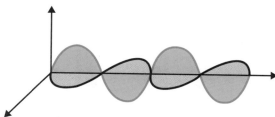

A An electromagnetic wave with oscillating electric and magnetic fields

B A mechanical wave moving in both transverse and longitudinal directions

C Motion of particles in a medium

D An electromagnetic wave with particles moving in both transverse and longitudinal directions

3 The diagram shows an experiment that was carried out to investigate the wave nature of light. What is the experiment and how did it support the wave theory of light?

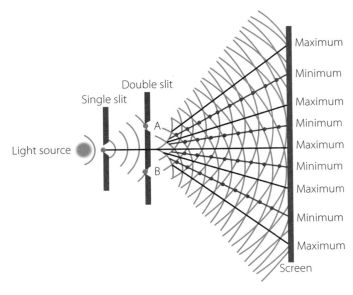

A Young's double-slit experiment, due to the light and dark bands seen on the screen. These indicate an interference pattern as seen in wave behaviour.

B The photoelectric effect, due to the light causing electrons to leave the surface of a metal.

C Foucault's rotating mirror, showing that the light will reflect back towards the source in a predictable wave like manner.

D Malus's polarisation of light experiment, due to the light being polarised along the x- and y-axes.

4 Two current-carrying wires are parallel and a distance of r metres apart, as shown in the diagram below. Wire 1 carries a current I_1 and wire 2 carries a current I_2. If $\vec{F}_{\text{by 2 on 1}}$ is the force that wire 2 exerts on wire 1 and $\vec{F}_{\text{by 1 on 2}}$ is the force that wire 1 exerts on wire 2, then it is true that:

A $\vec{F}_{\text{by 2 on 1}} = -\vec{F}_{\text{by 1 on 2}}$

B $\vec{F}_{\text{by 2 on 1}} = \vec{F}_{\text{by 1 on 2}}$

C No statement can be made comparing these forces because there is insufficient information about the currents in the wires.

D No statement can be made comparing these forces because there is insufficient information about the lengths of the wires.

9780170449687

5 An isotope of magnesium, Mg-23, is unstable and undergoes spontaneous β^+ decay. Which of the following shows the correct nuclear equation for the β^+ decay of Mg-23?

A $^{23}_{12}Mg \rightarrow ^{23}_{11}Na + ^{0}_{1}e + ^{0}_{0}\nu$

B $^{24}_{12}Mg \rightarrow ^{24}_{11}Na + ^{0}_{1}e + ^{0}_{0}\nu$

C $^{23}_{12}Mg \rightarrow ^{23}_{11}Na + ^{0}_{-1}e + ^{0}_{0}\nu$

D $^{23}_{12}Mg \rightarrow ^{23}_{13}Al + ^{0}_{-1}e + ^{0}_{0}\nu$

6 An object is swung in a circle horizontally as shown below.

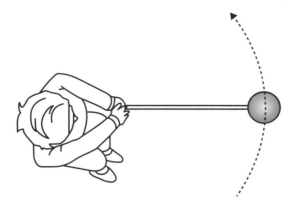

Which diagram below shows the path of the object if the person lets go at the point shown in the diagram above?

7 Monochromatic light of wavelength 425 nm is shone through a double slit as shown. The first maximum of the diffraction pattern is seen on a screen at an angle of 5° deflection from the central maximum. What is the distance between the slits?

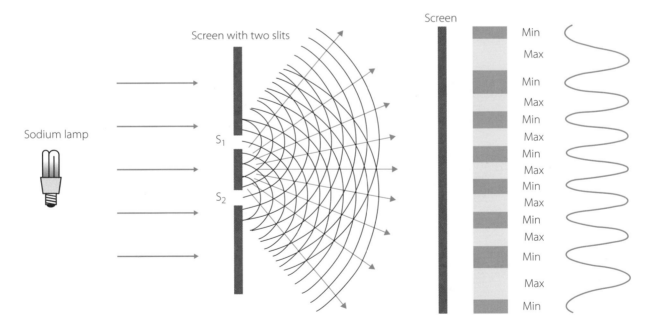

A 4.88×10^{-6} m

B 4.88×10^{-4} m

C 2.05×10^{3} m

D 2.05×10^{5} m

8 A 125 kg satellite is in orbit at an altitude of 630 km, as shown in the diagram. What is the speed of the satellite?

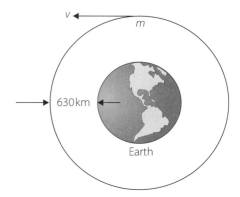

A $7560 \, \text{m s}^{-1}$

B $7925 \, \text{m s}^{-1}$

C $25\,204 \, \text{m s}^{-1}$

D $57\,153\,600 \, \text{m s}^{-1}$

9 If an electron in a hydrogen atom undergoes a transition from $n = 4$ to a lower energy level:

	Electron moves to $n = 3$	Electron moves to $n = 2$	Electron moves to $n = 1$
A	A photon is emitted of wavelength 1875 nm, which is in the infrared region of the spectrum.	A photon is emitted of wavelength 486 nm, which is in the visible region of the spectrum.	A photon is emitted of wavelength 97 nm, which is in the ultraviolet region of the spectrum.
B	A photon is emitted of wavelength 1875 nm, which is in the ultraviolet region of the spectrum.	A photon is emitted of wavelength 486 nm, which is in the visible region of the spectrum.	A photon is emitted of wavelength 97 nm, which is in the infrared region of the spectrum.
C	A photon is emitted of wavelength 97 nm, which is in the ultraviolet region of the spectrum.	A photon is emitted of wavelength 486 nm, which is in the visible region of the spectrum.	A photon is emitted of wavelength 1875 nm, which is in the infrared region of the spectrum.
D	A photon is emitted of wavelength 97 nm which is in the infrared region of the spectrum.	A photon is emitted of wavelength 486 nm, which is in the visible region of the spectrum.	A photon is emitted of wavelength 1875 nm, which is in the ultraviolet region of the spectrum.

10 Some motors are designed with magnets that are curved at the end as shown in the diagram below. This magnet shape results in a magnetic field that is described as 'radial' rather than uniform.

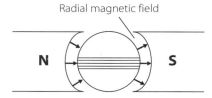

The purpose of designing a motor with a radial magnetic field rather than a uniform magnetic field is to:

A increase the force on the wire in the field as the current will be perpendicular to the magnetic field for a greater part of each rotation.

B increase the force on the wire in the field as the pivot arm will be perpendicular to the magnetic field for a greater part of each rotation.

C increase the torque on the wire in the field as the current will be perpendicular to the pivot arm for a greater part of each rotation.

D increase the torque on the coil in the field as the applied force will be perpendicular to the pivot arm for a greater part of each rotation.

11 A light was incident on a clean aluminium surface. A photoelectron was ejected when the photons of light had a minimum energy of $5.65 \times 10^{-19} \, \text{J}$.

What would be the maximum required wavelength required to emit the photoelectron?

A $8.53 \times 10^{14} \, \text{Hz}$

B $3.52 \times 10^{-7} \, \text{nm}$

C $3.52 \times 10^{-7} \, \text{m}$

D $8.53 \times 10^{14} \, \text{m}$

9780170449687

12 Three true statements about subatomic particles are given below.

- ▸ A lambda is a subatomic particle comprised of three quarks.
- ▸ A pion is a subatomic particle comprised of a quark and an antiquark.
- ▸ A tau is not a boson but always exists alone.

Use the statements to identify the category in which each subatomic particle belongs.

	Lambda	Pion	Tau
A	meson	baryon	lepton
B	boson	proton	hadron
C	baryon	lepton	meson
D	baryon	meson	lepton

13 Real transformers are not 100% efficient – they are not 'ideal'. In the analysis of ideal transformers, two assumptions are made. Which of the following is one of these assumptions?

A Lenz's Law does not apply to the primary coil.

B The core of the transformer is laminated – made out of many thin layers of soft iron, separated by insulating layers.

C Both primary and secondary coils consist of only a single loop of wire.

D The flux through the secondary coil is the same as that through the primary coil.

14 Which of the following options correctly identifies the evidence for and limitations of Rutherford's and Bohr's atomic models?

	Evidence for Rutherford's atomic model	Limitation of Rutherford's atomic model	Evidence for Bohr's atomic model	Limitation of Bohr's atomic model
A	Explained why electrons did not emit energy as they move in circular motion around the nucleus.	Unable to explain why the vast majority of alpha particles passed straight through the gold foil, but some were deflected at large angles.	Explained the differences in intensities of lines of emission spectra.	Unable to explain why each element had a unique emission spectrum featuring specific wavelengths of light.
B	The vast majority of alpha particles passed straight through the gold foil, but some were deflected at large angles.	Unable to explain why electrons did not emit energy as they move in circular motion around the nucleus.	Each element had a unique emission spectrum featuring specific wavelengths of light.	Unable to explain the differences in intensities of lines of emission spectra.
C	No alpha particles were able to pass through the gold foil.	Couldn't predict the charge of the nucleus.	Electrons existed in shells around the nucleus.	Used quantum physics to explain electron behaviour.
D	Each element had a unique emission spectrum featuring specific wavelengths of light.	Unable to explain the hyperfine lines of emission spectra of some elements.	The vast majority of alpha particles passed straight through the gold foil but some were deflected at large angles.	Unable to explain why electrons did not emit energy as they move in circular motion around the nucleus.

15 Two children are on a solid plank that is balanced on a pivot point. The child on the left has a mass of 31 kg.

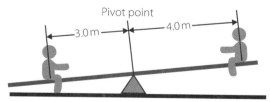

What mass of the child on the right would allow the plank to be balanced horizontally?

A 4.43 kg

B 23.25 kg

C 30.00 kg

D 41.33 kg

16 An unstable isotope of potassium, K-40, is used for the radiometric dating of rocks. By looking at how much of the isotope is present in the rock sample, the age of the rock can be determined. K-40 has a decay constant of 5.55×10^{-10} year^{-1}. Determine the most appropriate value for the half-life of K-40.

   A 2.62 billion years

   B 1.25 billion seconds

   C 3.85×10^{-10} seconds

   D 1.26 billion years

17 When cosmic rays emitted by stellar objects encounter Earth's atmosphere, this results in particles decaying or new particles being created. One of the particles that can be created is the muon. Experiments on Earth indicate that the half-life of a stationary muon is 2.1 μs and that muons created in the atmosphere travel near to the speed of light (0.994c).

According to Newtonian physics, it can be shown that a muon will travel nearly 626 m in its 2.1 μs lifetime; however, muons are detected at Earth's surface. Which row of the table correctly indicates how special relativity can explain muons reaching Earth's surface?

	Frame of reference – Earth		Frame of reference – Muon	
	Distance travelled by muon (m)	Lifetime of muon (μs)	Distance travelled by muon (m)	Lifetime of muon (μs)
A	626	2.1	>626	>2.1
B	>626	>2.1	626	2.1
C	<626	<2.1	626	2.1
D	626	2.1	<626	<2.1

18 The diagram below represents four of the moons in orbit around Jupiter. Callisto has an orbital radius four times that of Io. What would be the ratio of the orbital period of Callisto to the orbital period of Io?

Callisto   Ganymede   Europa   Io   Jupiter

Not to scale

   A 1:8

   B 8:1

   C 22.63:1

   D 1:22.63

9780170449687

19 A stationary electron is released in an electric field such as shown below. It accelerates while in the electric field and reaches a velocity v.

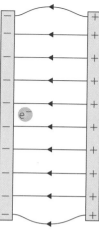

The voltage V between the plates is varied and data collected about how this change affects the final velocity of the electron in the electric field. If the data is plotted, which graph will yield a straight line?

A V vs v

B V vs v^2

C V^2 vs v

D V vs $\dfrac{1}{v}$

20 Which of the following particles would give the smallest radius of orbit if all were injected at the same velocity into a region with a magnetic field perpendicular to the initial velocity, as shown below?

$$
\begin{array}{ccccc}
\times & \times & \times & \times & \times \\
\times & \times & \times & \times & \times \\
\times & \times & \times & \times & \times \quad \longleftarrow \bullet \\
\times & \times & \times & \times & \times \\
\times & \times & \times & \times & \times \\
\end{array}
$$

A neutron

B proton

C alpha particle

D photon

Section 2

80 marks

Question 21 (5 marks)

A lemon is launched upwards from the edge of a cliff, at $40\,\mathrm{m\,s^{-1}}$, at an angle of $20°$ above the horizontal, and lands in the ocean 70 m below.

  a When is the lemon moving slowest during its trajectory? Justify your response.   (2 marks)

_____

_____

_____

_____

  b Calculate the greatest speed the lemon has during its trajectory.   (3 marks)

_____

_____

_____

_____

Question 22 (6 marks)

Two particles of equal mass are fired into a uniform magnetic field at equal speeds. They then follow the paths shown in the diagram below.

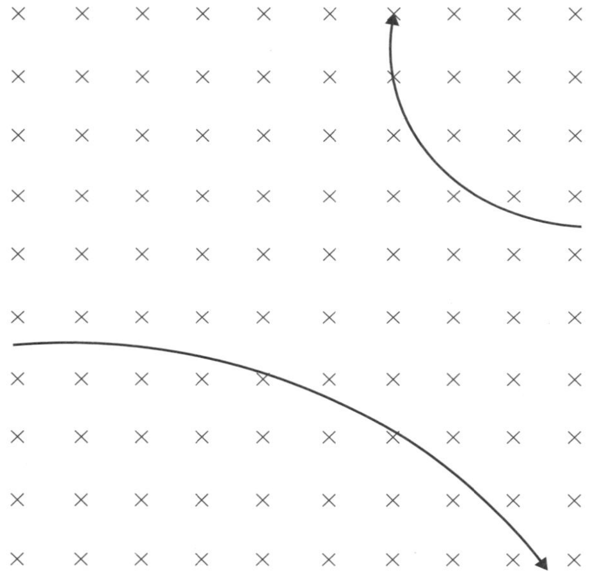

  a Explain a characteristic of the particles that **must** be the same and one that **must** be different from the information contained in the diagram.   (4 marks)

_____

_____

_____

  b Account for the shape of the path of one of the particles.   (2 marks)

_____

_____

Question 23 (6 marks)

A student measures the energy of photoelectrons emitted from a metal when light of different frequencies is shone onto it. The data collected is shown below.

Frequency ($\times 10^{14}$ Hz)	Energy of the photoelectron ($\times 10^{-19}$ J)
6.5	1.18
8.4	1.60
9.4	3.40
10	3.04
10.5	3.35
11.6	4.00

By plotting an appropriate graph on the axes below, determine the work function and cut-off frequency of the metal. Show all your working on the graph.

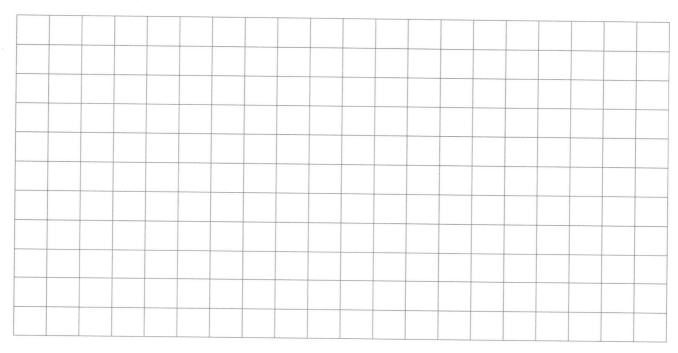

Question 24 (4 marks)

A conductor, as seen below, is placed in a magnetic field (B) so that 35 cm of its length is within the field. The magnetic field intensity is 25 mT.

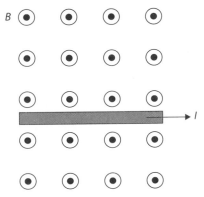

a   Determine the force experienced by the conductor when a current of 0.40 A passes through it in the direction shown. (2 marks)

_____

_____

b   Explain any changes in the magnitude of the force experienced by the conductor as it is turned through 90° so that the current now flows directly down the page. (2 marks)

_____

_____

Question 25 (4 marks)

The orbital velocity for a planet or satellite can be found using information about the centripetal force acting on the orbiting object. The centripetal force is provided by the gravitational field of the object at the centre of the circle.

a   Derive an equation for orbital velocity in terms of the gravitational constant, G, the mass of the orbited object, M, and the orbital radius, r. (2 marks)

_____

_____

_____

b   Given Earth has a mass of 6.0×10^{24} kg and a radius of 6370 km, determine the speed a satellite would need to move with to be in a stable orbit at an altitude of 1000 km. (2 marks)

_____

_____

_____

9780170449687

Question 26 (10 marks)

Stars are powered by nuclear reactions in their cores. For any given star, the type and location of these reactions are indicated by the star's position on the Hertzsprung–Russell diagram. Distinct groups can be seen on the Hertzsprung–Russell diagram below. Stars V, W, X, Y and Z are indicated on the diagram within those groups.

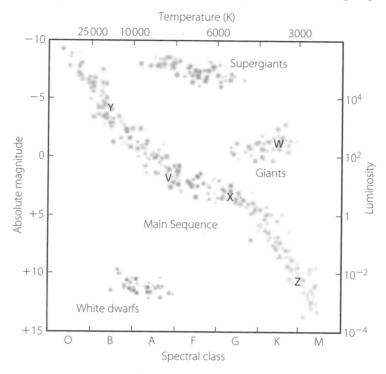

a Compare the fusion reactions occurring in the core of stars Y and Z in the diagram above.   (4 marks)

_____

_____

_____

_____

_____

_____

_____

_____

_____

_____

_____

_____

_____

_____

_____

_____

_____

_____

b Outline the differences in the fusion reactions of X and W. (3 marks)

_____

_____

_____

_____

_____

_____

c With reference to at least two evolutionary stages, explain the link between the colour and luminosity of star V. (3 marks)

_____

_____

_____

_____

9780170449687

Question 27 (5 marks)

An investigation is performed using a transformer with 500 turns on the primary coil. The primary voltage is changed and the secondary voltage measured. Collected data is below.

a Use the data to plot a graph on the axes provided that will enable you to determine the number of turns on the secondary coil.

(4 marks)

V_P (V)	V_S (V)
2.0	0.30
3.9	0.65
5.9	0.95
8.0	1.25
10.1	1.60
11.9	1.90

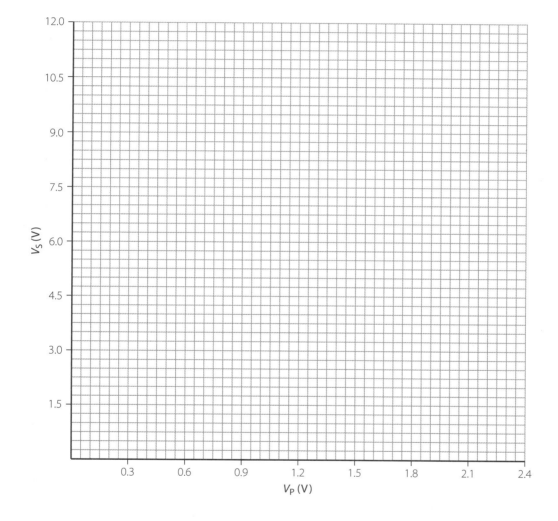

_____

_____

_____

b Make a statement about the reliability of the data collected.

(1 mark)

_____

_____

Question 28 (4 marks)

Luke ties a sturdy rope to a heavy, smooth metal disc. Luke moves the metal disc in uniform circular motion through two circles of identical radius, at identical speed – one circle with the rope horizontal and the disc on a frictionless surface, and the other circle with the rope vertical. Compare the centripetal force on the disc and tension force in the rope in each situation.

_____

_____

_____

_____

_____

_____

_____

_____

_____

_____

Question 29 (6 marks)

The graphic below can be used to describe the sequence of the scientific method.

With reference to the scientific method, evaluate the evidence that supports Einstein's Theory of Special Relativity.

_____

_____

_____

_____

_____

_____

_____

_____

_____

_____

_____

_____

_____

_____

_____

_____

_____

_____

_____

_____

Question 30 (9 marks)

U-235 is capable of undergoing nuclear fission according to the following equation.

$$^{235}_{92}U + {}^{1}_{0}n \rightarrow {}^{144}_{56}Ba + {}^{89}_{36}Kr + 3\,{}^{1}_{0}n$$

a A nuclear chain reaction occurs when product neutrons cause subsequent nuclear fission reactions. With reference to the equation above, compare the requirements of a controlled and an uncontrolled nuclear reaction. (3 marks)

_____

_____

_____

_____

_____

_____

_____

_____

b Calculate the energy released when 15.45 kg of U-235 undergoes fission, given that the fission of a single atom of U-235 releases 205 MeV and there are 2.563×10^{21} U-235 atoms per kg. Provide your answer in joules. (2 marks)

_____

_____

c U-238 naturally undergoes alpha decay.

  i Write a nuclear equation for this decay. (2 marks)

_____

  ii Explain how alpha decay can lead to the formation of more stable radioisotopes. (2 marks)

_____

_____

_____

_____

Question 31 (7 marks)

Electrons in uniform gravitational fields and in uniform electric fields are subject to forces and, thus, have a specific and predictable motion when moving within the fields.

The diagram below shows electrons being fired horizontally at $4.32 \times 10^6 \, \text{m s}^{-1}$ into a vertically oriented electric field (of strength 55000 V/m) and a uniform gravitational field at the surface of Earth. The motion of the electron is initially perpendicular to the field in each case.

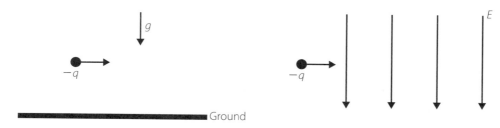

Compare qualitatively and quantitatively the motion of the electrons in the two fields.

_____

_____

_____

_____

_____

_____

_____

_____

_____

_____

_____

_____

Question 32 (5 marks)

With reference to the diagram below, explain the process by which torque is produced in an AC induction motor.

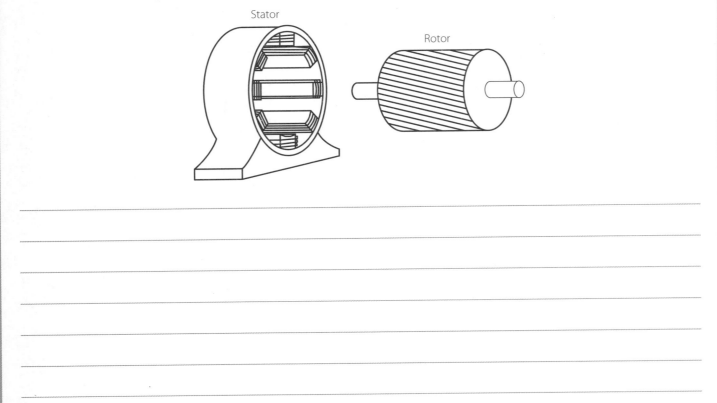

Stator

Rotor

_____

_____

_____

_____

_____

_____

_____

9780170449687

Question 33 (9 marks)

Wave–particle duality, as proposed by Einstein, is not limited to light or other forms of electromagnetic radiation. De Broglie proposed that particles could have wave properties. It is said that 'if sufficient evidence supports a hypothesis then it can be considered a valid theory'. Evaluate the validity of the theory proposed by each scientist. Include one piece of experimental evidence that supported the theory and describe the implications of each theory in the development of the respective scientific models.

Key

Symbol of element:

- □ gas at room temperature
- □ liquid at room temperature
- ■ solid at room temperature
- □ synthetic (does not occur naturally)

atomic number → 26

Fe

name of element → iron

standard atomic weight → 55.85

- □ s block
- □ p block
- □ d block transition metals
- □ d block lanthanoids and actinoids

1																	18
1 **H** hydrogen 1.008	2											13	14	15	16	17	**2** **He** helium 4.003
3 **Li** lithium 6.941	**4** **Be** beryllium 9.012											**5** **B** boron 10.81	**6** **C** carbon 12.01	**7** **N** nitrogen 14.01	**8** **O** oxygen 16.00	**9** **F** fluorine 19.00	**10** **Ne** neon 20.18
11 **Na** sodium 22.99	**12** **Mg** magnesium 24.31	3	4	5	6	7	8	9	10	11	12	**13** **Al** aluminium 26.98	**14** **Si** silicon 28.09	**15** **P** phosphorus 30.97	**16** **S** sulfur 32.07	**17** **Cl** chlorine 35.45	**18** **Ar** argon 39.95
19 **K** potassium 39.10	**20** **Ca** calcium 40.08	**21** **Sc** scandium 44.96	**22** **Ti** titanium 47.87	**23** **V** vanadium 50.94	**24** **Cr** chromium 52.00	**25** **Mn** manganese 54.94	**26** **Fe** iron 55.85	**27** **Co** cobalt 58.93	**28** **Ni** nickel 58.69	**29** **Cu** copper 63.55	**30** **Zn** zinc 65.38	**31** **Ga** gallium 69.72	**32** **Ge** germanium 72.63	**33** **As** arsenic 74.92	**34** **Se** selenium 78.96	**35** **Br** bromine 79.90	**36** **Kr** krypton 83.80
37 **Rb** rubidium 85.47	**38** **Sr** strontium 87.61	**39** **Y** yttrium 88.91	**40** **Zr** zirconium 91.22	**41** **Nb** niobium 92.91	**42** **Mo** molybdenum 95.96	**43** **Tc** technetium	**44** **Ru** ruthenium 101.1	**45** **Rh** rhodium 102.9	**46** **Pd** palladium 106.4	**47** **Ag** silver 107.9	**48** **Cd** cadmium 112.4	**49** **In** indium 114.8	**50** **Sn** tin 118.7	**51** **Sb** antimony 121.8	**52** **Te** tellurium 127.6	**53** **I** iodine 126.9	**54** **Xe** xenon 131.3
55 **Cs** caesium 132.9	**56** **Ba** barium 137.3	57–71 lanthanoids	**72** **Hf** hafnium 178.5	**73** **Ta** tantalum 180.9	**74** **W** tungsten 183.9	**75** **Re** rhenium 186.2	**76** **Os** osmium 190.2	**77** **Ir** iridium 192.2	**78** **Pt** platinum 195.1	**79** **Au** gold 197.0	**80** **Hg** mercury 200.6	**81** **Tl** thallium 204.4	**82** **Pb** lead 207.2	**83** **Bi** bismuth 209.0	**84** **Po** polonium	**85** **At** astatine	**86** **Rn** radon
87 **Fr** francium	**88** **Ra** radium	89–103 actinoids	**104** **Rf** rutherfordium	**105** **Db** dubnium	**106** **Sg** seaborgium	**107** **Bh** bohrium	**108** **Hs** hassium	**109** **Mt** meitnerium	**110** **Ds** darmstadtium	**111** **Rg** roentgenium	**112** **Cn** copernicium	**113** **Nh** nihonium	**114** **Fl** flerovium	**115** **Mc** moscovium	**116** **Lv** livermorium	**117** **Ts** tennessine	**118** **Og** oganesson

57 **La** lanthanum 138.9	**58** **Ce** cerium 140.1	**59** **Pr** praseodymium 140.9	**60** **Nd** neodymium 144.2	**61** **Pm** promethium	**62** **Sm** samarium 150.4	**63** **Eu** europium 152.0	**64** **Gd** gadolinium 157.3	**65** **Tb** terbium 158.9	**66** **Dy** dysprosium 162.5	**67** **Ho** holmium 164.9	**68** **Er** erbium 167.3	**69** **Tm** thulium 168.9	**70** **Yb** ytterbium 173.1	**71** **Lu** lutetium 175.0
89 **Ac** actinium	**90** **Th** thorium 232.0	**91** **Pa** protactinium 231.0	**92** **U** uranium 238.0	**93** **Np** neptunium	**94** **Pu** plutonium	**95** **Am** americium	**96** **Cm** curium	**97** **Bk** berkelium	**98** **Cf** californium	**99** **Es** einsteinium	**100** **Fm** fermium	**101** **Md** mendelevium	**102** **No** nobelium	**103** **Lr** lawrencium

Fully worked solutions are provided below to demonstrate the steps necessary to reach the required answer. Worked solutions help you independently review your own answers.

MODULE FIVE: ADVANCED MECHANICS

REVIEWING PRIOR KNOWLEDGE PAGE 1

1   a   Distance = speed × time
$$= 60 \text{ km h} \times 1.5 \text{ h}$$
$$= 90 \text{ km} = 90\,000 \text{ m}$$

   b $\text{Time} = \dfrac{\text{Distance}}{\text{Speed}}$

$$= \dfrac{400 \text{ m}}{5 \text{ m s}^{-1}} = 80 \text{ seconds}$$

   c $\text{Speed} = \dfrac{\text{Distance}}{\text{Time}}$

$$= \dfrac{25\,000 \text{ m}}{4000 \text{ s}} = 6.25 \text{ m s}^{-1} = 6.3 \text{ m s}^{-1}$$

2   a   Acceleration is the change of speed or direction, or both, of an object. These collectively can be described as a change of velocity.

   b   i   The rubber ball undergoes acceleration because it changes direction.

      ii   The train does not undergo acceleration because it travels at constant speed in one direction.

      iii   The car undergoes acceleration because it changes direction.

      iv   The car undergoes acceleration because it will change speed.

3   a $a = \dfrac{v - u}{t}$

   b $u = \sqrt{v^2 - 2as}$

   c $t = \sqrt{\dfrac{2s}{a}}$

4   a $v = u + at$
$$4 = 60 + -1.4t$$
$$t = 40 \text{ s}$$

   b $s = ut + \dfrac{1}{2}at^2$
$$s = 60 \times 40 + \dfrac{1}{2} \times -1.4 \times 40^2$$
$$s = 1280 \text{ m} = 1300 \text{ m}$$

5   a $F_{net} = m_{total} \times a$ (Newton's second law)
$$40 = 16 \times a$$
$$a = 2.5 \text{ m s}^{-2}$$

   b   Block B accelerates at 2.5 m s^{-2}.
$$F_{net} = m_{total} \times a \text{ (Newton's second law)}$$
$$F_{A \text{ on } B} = F_{net \text{ on } B}$$
$$= m_B \times a_B$$
$$= 10 \times 2.5 = 25 \text{ N right}$$

   c $F_{A \text{ on } B} = -F_{B \text{ on } A}$
Therefore, $F_{B \text{ on } A} = 25 \text{ N left}$
This is a consequence of Newton's third law. The force exerted on block A by block B must be equal and opposite to the force exerted by block A on block B.

6   a $F_x = F \times \cos \theta = 150 \times \cos 25° = 135.946 = 136 \text{ N}$ (Note that significant figures of angles are normally not considered at this level.)
$$F_y = F \times \sin \theta$$
$$= 150 \times \sin 25° = 63.3927 = 63.4 \text{ N}$$

   b   i $W = Fs = 150 \times 1.2 = 200 \text{ J}$

      ii $U_{gained} = W_{done} = 200 \text{ J}$

      iii   For a falling object (assuming no air resistance), $K_{gained} = U_{lost} = 200 \text{ J}$

Chapter 1: Projectile motion

WS 1.1 PAGE 4

1   This shows that the vertical motion of projectiles (both bullets in this instance) is independent of their horizontal motion.

2   The shadow represents only the horizontal component of the motion. Thus, the motion of the shadow is constant because horizontal motion is not accelerated motion.

3   a   The apple will land next to the mast, directly under the point from which it was released.

   b   Philip will see the apple move in a parabolic arc as it continues to move forward at the same constant speed as the boat as it falls (accelerates) down.

4   a   Velocity north $= u_x = u \cos \theta$
$$= 100 \cos 75.0° = 25.9 \text{ m s}^{-1}$$

   b   Velocity up $= u_y = u \sin \theta$
$$= 100 \sin 75.0° = 96.6 \text{ m s}^{-1}$$

5   a $v_y = 0$ at maximum height because the projectile has stopped going up and is yet to start moving down.

   b   Consequently, at this point the projectile will have its lowest velocity value because its total velocity will be v_x. At all other points there will be a non-zero v_y that will add to the total velocity ($v_{total}^2 = v_x^2 + v_y^2$).

6   $s_x = u_x t + \dfrac{1}{2}at^2$
Since $a = 0$ then $s_x = u_x t$
$u_x = u \cos \theta$; thus $s_x = u \cos \theta \, t$ and is hence dependent only on u, θ and t.

7

	Up	Down
Vertical speed	$v_{up}^2 = u_{up}^2 + 2as_{up}$ and, because $v_{up} = 0$, $$0 = u_{up}^2 + 2 \times -g \times s_{up}$$ $$u_{up}^2 = 2gs_{up}$$	$v_{down}^2 = u_{down}^2 + 2as_{down}$ and, because $u_{down} = 0$, $$v_{down}^2 = 2 \times -g \times -s_{down}$$ so $v_{down}^2 = 2gs_{down}$ $$= u_{up}^2$$ so $v_{down} = -u_{up}$
Time	$v_{up} = u_{up} + at_{up}$ and, because $v_{up} = 0$, $$0 = u_{up} + at_{up}$$ $t_{up} = \dfrac{-u_{up}}{-g}$ This then equals $\dfrac{u_{up}}{g}$	$v_{down} = u_{down} + at_{down}$ and, because $u_{down} = 0$, $$v_{down} = at_{down}$$ $$t_{down} = \dfrac{v_{down}}{-g}$$ so if $v_{down} = -u_{up}$ then $t_{up} = t_{down}$

WS 1.2 PAGE 6

1 a $s_y = 0$, $u_y = 10 \times \sin 30° \, \text{m s}^{-1}$, $a = -9.8 \, \text{m s}^{-2}$

$$s_y = u_y t + \frac{1}{2}at^2$$
$$0 = 5t - 4.9t^2 = t(5 - 4.9t)$$
$$t = 0, \ 1.020408 \, \text{s}$$

Thus, time of flight = 1.0 s.

b Maximum height $s_{y\,max}$ is reached at $t = \dfrac{5}{9.8} \, \text{s}$ (from part **a**),

$u_y = 10 \times \sin 30° \, \text{m s}^{-1}$, $a = -9.8 \, \text{m s}^{-2}$

~continued in right column ▲

$$s_{y\,max} = u_y t + \frac{1}{2}at^2$$
$$= 5 \times \frac{5}{9.8} - 4.9 \times \left(\frac{5}{9.8}\right)^2$$
$$= 1.2755 = 1.3 \, \text{m}$$

c $u_x = 10\cos 30° \, \text{m s}^{-1}$, $t = \dfrac{5}{4.9} \, \text{s}$

Range $= s_x = u_x \times t = 8.83699 = 8.8 \, \text{m}$

d Final speed $v = u = 10 \, \text{m s}^{-1}$ (symmetry of motion on flat ground)

2 By substituting in $u = 20 \, \text{m s}^{-1}$ and then $u = 30 \, \text{m s}^{-1}$ and following the same calculations, the following values are obtained.

	Quantity	$u = 10 \, \text{m s}^{-1}$	$u = 20 \, \text{m s}^{-1}$	$u = 30 \, \text{m s}^{-1}$
a	Time of flight	1.0 s	2.0 s	3.1 s
b	Maximum height	1.3 m	5.1 m	12 m
c	Range	8.8 m	35 m	79 m
d	Final speed	10 m s^{-1}	20 m s^{-1}	30 m s^{-1}

Allowing for rounding, the following relationships can be noted:

a Time of flight is proportional to launch speed.

b Maximum height is proportional to (launch speed)^2.

c Range is proportional to (launch speed)^2.

d Final speed is equal to launch speed.

$$s_y = u_y t + \frac{1}{2} at^2$$
$$0 = u\sin\theta \, t - 4.9t^2$$
$$t = 0, \ \frac{u\sin\theta}{4.9} \, \text{s}$$

Therefore, time of flight proportional to $\sin\theta$.

3 a

θ (°)	15	30	45	60	75
t (s)	5.2	10	14	17	19

Sketch graph is a sin curve:

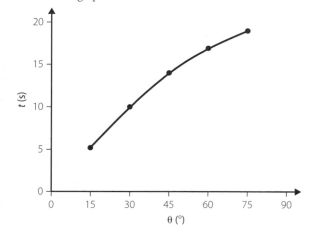

b

θ (°)	15	30	45	60	75
v_y (m s^{-1})	25	49	69	85	94

Sketch graph is a sin curve.

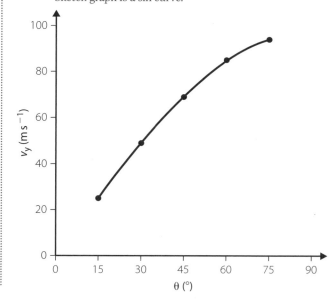

9780170449687

$v_y = u_y + at$
$\quad = u\sin\theta - 9.8t = u\sin\theta - 9.8 \times \dfrac{u\sin\theta}{4.9}$
$\quad = -u\sin\theta$

Therefore, final vertical velocity proportional to $\sin\theta$.

c

θ (°)	15	30	45	60	75
s_x (m)	490	850	980	850	490

Sketch graph is a curve of sinusoidal shape with a maximum value at $\theta = 45°$.

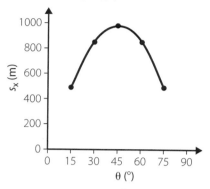

$s_x = u_x t = u\cos\theta \times \dfrac{u\sin\theta}{4.9} = \dfrac{u^2\sin 2\theta}{9.8}$ (not expected knowledge)

Therefore, range is proportional to $\cos\theta \times \sin\theta$ or $\sin 2\theta$.

~continued in right column ▲

d

θ (°)	15	30	45	60	75
s_y (m)	33	120	250	370	460

Sketch graph is a curve of sinusoidal shape.

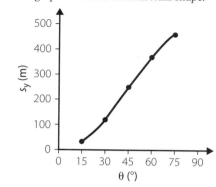

$s_y = u_y t + \dfrac{1}{2}at^2$ (where $t = \dfrac{\text{time of flight}}{2} = \dfrac{u\sin\theta}{9.8}$)

$\quad = u\sin\theta \times \dfrac{u\sin\theta}{9.8} - 4.9 \times \left(\dfrac{u\sin\theta}{9.8}\right)^2$

$\quad = \dfrac{u^2\sin^2\theta}{19.6}$

Therefore, maximum height is proportional to $\sin^2\theta$.

---

4 $u_x = 50\cos 40°$, $u_y = 50\sin 40°$, $s_y = 0$, $a_{\text{Earth}} = -9.8\,\text{m s}^{-2}$, $a_{\text{Moon}} = -1.6\,\text{m s}^{-2}$

Quantity	Earth	Moon
Time of flight	$s_y = u_y t + \dfrac{1}{2}at^2$ $0 = 50\sin 40° \times t - 4.9t^2$ $t = 6.6\,\text{s}$	$s_y = u_y t + \dfrac{1}{2}at^2$ $0 = 50\sin 40° \times t - 0.8t^2$ $t = 40\,\text{s}$
Range	$s_x = u_x \times t$ $\quad = 50\cos 40° \times 6.56$ $\quad = 250\,\text{m}$	$s_x = u_x \times t$ $\quad = 50\cos 40° \times 40.2$ $\quad = 1500\,\text{m}$
Maximum height	$s_y = u_y t + \dfrac{1}{2}at^2$ (t = half time of flight) $s_y = 50\sin 40° \times 3.28 - 4.9 \times 3.28^2$ $\quad = 53\,\text{m}$	$s_y = u_y t + \dfrac{1}{2}at^2$ (t = half time of flight) $s_y = 50\sin 40° \times 20.1 - 0.8 \times 20.1^2$ $\quad = 320\,\text{m}$

All three quantities – time of flight, range, maximum height – are proportionally greater on the Moon than on Earth, by a factor equivalent to the ratio of gravitational acceleration.

5 a Minimum launch speed means vertical launch and $v_y = 0$ at top.
$v_y = 0$, $a = -9.8\,\text{m s}^{-2}$, $s_y = 18\,\text{m}$
$v_y^2 = u_y^2 + 2as_y$
$0 = u_y^2 + 2 \times -9.8 \times 18$
$u_y = 19\,\text{m s}^{-1}$

b i Final velocity will comprise both vertical and horizontal components.
$u_y = 15\sin 10°$, $a = -9.8\,\text{m s}^{-2}$, $s_y = -18\,\text{m}$
$v_y^2 = u_y^2 + 2as_y$
$\quad = (15\sin 10°)^2 + 2 \times -9.8 \times -18$
Therefore $v_y = 18.601\,489 = 19\,\text{m s}^{-1}$ down
$u_x = 15\cos 10°$
Therefore $v_x = 15\cos 10°$
$v^2 = v_y^2 + v_x^2$
$v = \sqrt{(18.601\,489)^2 + (15\cos 10°)^2}$
$\quad = 23.753535 = 24\,\text{m s}^{-1}$

To find s_x we first need to calculate time of flight.
$$v_y = u_y + at$$
$$-19 = 15\sin 10° + -9.8 \times t$$
$$t = 2.20456\,\text{s}$$
$$s_x = u_x t = 15\cos 10° \times 2.20456 = 33\,\text{m}$$

6 a $a = -9.8\,\text{m s}^{-2}$, $s_y = -800\,\text{m}$, $u = 12\,\text{m s}^{-1}$, $\theta = 15°$
$u_y = 12\sin 15°$ down $= -3.1\,\text{m s}^{-1}$
Finding v_y to find t:
$$v_y^2 = u_y^2 + 2as_y = (12\sin 15°)^2 + -19.6 \times -800$$
$$v_y = 125.2583\,\text{m s}^{-1}\ \text{down}$$
$$v_y = u_y + at$$
$$-125.2583 = -3.1 + -9.8t$$
$$t = 12.46513\,\text{s}$$
Distance $= s_x = u_x t = 12\cos 15° \times 12.46513 = 140\,\text{m}$

b Finding speed: $v^2 = v_y^2 + v_x^2 = 125.2583^2 + (12\cos 15)^2$
$$= 125.7852 = 130\,\text{m s}^{-1}.$$

7 $a = -9.8\,\text{m s}^{-2}$, $s_y = 0\,\text{m}$ (caught at same height),
$u = 25\,\text{m s}^{-1}$, $\theta = 60°$
$u_y = 25\sin 60° = -v_y$ (symmetrical nature of trajectory)
Find t:
$$v_y = u_y + at$$
$$-25\sin 60° = 25\sin 60° + -9.8t$$
$$t = 4.418497\,\text{s}$$
Range $= s_x = u_x t$
$$= 25\cos 60° \times 4.418497 = 55.23121\,\text{m}$$
Ellyse runs a distance of $90 - 55.23121$ metres in a time of $4.4218497\,\text{s}$
So speed $= v = \dfrac{s}{t} = \dfrac{34.768787}{4.4218497} = 7.9\,\text{m s}^{-1}$

Chapter 2: Circular motion

WS 2.1 PAGE 11

1 The tension force in the rope is providing the necessary centripetal force. Normally an ice-skater would have used a force generated between the edge of the skates and the ice to provide F_c.

2 a Friction between the coin and the rubber turntable surface provides the centripetal force.

b In both cases, the centripetal force is a frictional force. For the record-player, this occurs between the turntable and the coin; for the car and bicycle, this occurs between the tyres and the road.

c Friction between two surfaces is a function of the mass of the object and a coefficient determined by those surfaces. In the cases discussed, the mass and coefficient will be constant and, therefore, the frictional force will be constant. If the speed increases, more force is required to sustain circular motion because $F_c = \dfrac{mv^2}{r}$. If the force required exceeds that which can be provided by friction the objects will no longer continue in circular motion.

3 The attractive electrostatic force acting between the oppositely charged proton and electron is providing the necessary centripetal force. The F_c is directed at the proton. This F_c is a function of the magnitude of the two charges and the square of the distance between them.

4 The attractive gravitational force acting between the satellite and Earth is providing the necessary centripetal force. The F_c is directed at the centre of Earth. This F_c is a function of the magnitude of the two masses and the square of the distance between their centres.

5 The normal force acting on the people from the wall is providing the necessary centripetal force directed towards the centre of the circle. This F_c is a function of the speed of circular motion: as they rotate faster the normal force will increase, thus acting to push them onto the wall. Since friction is a function of the normal force, friction will also be increasing as they rotate faster. At some point the frictional force will be sufficient to prevent people sliding down and so the floor is no longer required to stop people from falling.

6

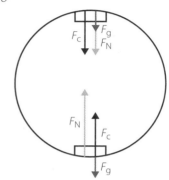

The centripetal force must be constant for uniform circular motion. There are two forces acting on the rollercoaster – a normal force (F_N) from the rails and the gravitational force (F_g). The gravitational force is directed down throughout the motion, while the normal force is always towards the centre of the loop. For F_c to be of constant magnitude, the normal force must be largest at the bottom of the loop (when F_g is in the opposite direction) and smallest at the top (when F_g is in the same direction).

7

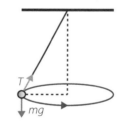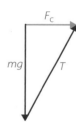

The necessary centripetal force is provided by the vector sum of the tension force (T) acting along the string and the gravitational force (mg) on the mass.

8 The vector sum of the normal force (F_N) acting perpendicular to the surface and gravitational force (mg) on the car provides the additional centripetal force. Alternatively, this can be analysed as the horizontal component of the normal force providing the centripetal force while the vertical component is equal and opposite to the gravitational force. These two analyses are equivalent.

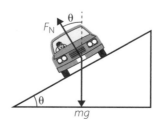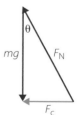

9780170449687

WS 2.2 PAGE 15

1. $a_c = \dfrac{v^2}{r} = \dfrac{12.0^2}{10.0} = 14.4\,\text{m s}^{-2}$

2. If the coin does 33 revolutions in one minute then the period of revolution is 1.81 s.

   Then $v = \dfrac{2\pi r}{T} = 2\pi \times \dfrac{0.1}{1.81}$
   $= 0.347\,137 = 0.35\,\text{m s}^{-1}$

3. One billion revolutions $= 2 \times 10^9\,\pi$ radians.

   $\omega = \dfrac{\Delta\theta}{t}$, so angular velocity $= 4.11 \times 10^{16} = \dfrac{2 \times 10^9\,\pi}{t}$

   Thus, $t = 1.528\,755 \times 10^{-7} = 1.53 \times 10^{-7}\,\text{s}$

4. a $r = (6370 + 100) \times 10^3\,\text{m}$

   so $a_c = \dfrac{v^2}{r} = \dfrac{7850^2}{6\,470\,000}$
   $= 9.524\,343 = 9.52\,\text{m s}^{-2}$

   b $v = \dfrac{2\pi r}{T}$ so $T = \dfrac{2\pi r}{v} = \dfrac{2\pi \times 6\,470\,000}{7850}$
   $= 5178.6\,\text{s} = 86$ minutes and 19 seconds.

   c $\omega = \dfrac{\Delta\theta}{t} = \dfrac{2\pi}{5178.6}$
   $= 1.213\,439 \times 10^{-3} = 1.21 \times 10^{-3}\,\text{radian s}^{-1}$

   It can be seen that multiplying this value by the radius yields the speed of the satellite and, hence, $v = r\omega$.

5. $F_{\text{friction}} = \mu F_N = mg$ and $F_N = F_c = \dfrac{mv^2}{r}$

   Therefore $\dfrac{\mu m v^2}{r} = mg$

   Thus $\mu v^2 = rg$

   $\mu = \dfrac{rg}{v^2} = 5 \times \dfrac{9.8}{8.5^2}$

   $\mu = 0.678 = 0.68$

6. At the minimum speed, the centripetal force will be entirely provided by force due to gravity (mg) for uniform circular motion.

   $F_c = \dfrac{mv^2}{r} = mg$

   Therefore $v^2 = rg = 30 \times 9.8$

   So $v = 4.9\,\text{m s}^{-1}$

7. a $\sin\theta = \dfrac{r}{L} = \dfrac{0.10}{1.05}$
   so $\theta = 5.465 = 5.5°$

   b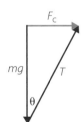

   $\tan\theta = \dfrac{\left(\dfrac{mv^2}{r}\right)}{mg} = \dfrac{v^2}{rg}$

   Thus $v^2 = rg\tan\theta = 0.10 \times 9.8 \times \tan 5.465° = 0.093\,76$

   Thus $v = 0.3062 = 0.31\,\text{m s}^{-1}$

   c $T = \dfrac{2\pi r}{v} = \dfrac{2\pi \times 0.1}{0.31} = 2.051\,98 = 2.0\,\text{s}$

8. The vector sum of the normal force (F_N) acting perpendicular to the surface and force due to gravity on the car (mg) provides the required centripetal force.

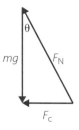

   $108\,\text{km h}^{-1} = 30.0\,\text{m s}^{-1}$

   $\tan\theta = \dfrac{\left(\dfrac{mv^2}{r}\right)}{mg} = \dfrac{v^2}{rg} = \dfrac{30.0^2}{1100 \times 9.8}$

   Thus $\theta = 4.77°$

WS 2.3 PAGE 18

1. a The straight line of best fit indicates that torque and force are proportional.

   b The collected data are not reliable. There is significant scatter of the points about the line of best fit, indicating poor reliability.

   c The data collected is of good accuracy. The line of best fit extrapolates through (0, 0) on the graph, which would be expected from the relationship between these variables indicated by the equation.

   d The gradient of the line of best fit represents $\dfrac{\text{rise}}{\text{run}}$ for the plotted data. This is equivalent to $\dfrac{\tau}{F}$. From the equation it can be seen that $\dfrac{\tau}{F} = r\sin\theta$. Since the gradient is approximately $\dfrac{15 - 0}{20 - 0} = 0.75$, it can be stated that $r\sin\theta = 0.75$.

2. a The straight line of best fit indicates that τ and r are proportional.

   b The collected data are reliable. The scatter of the points about the line of best fit is minimal, indicating good reliability.

   c The data collected are of poor accuracy. The line of best fit does not extrapolate through or close to (0, 0), which is contrary to what would be expected from the relationship between these variables indicated by the equation.

   d The gradient of the line of best fit represents $\dfrac{\text{rise}}{\text{run}}$ for the plotted data. This is equivalent to $\dfrac{\tau}{r}$. From the equation it can be seen that $\dfrac{\tau}{r} = F\sin\theta$. Since the gradient is approximately $\dfrac{10 - 3.8}{1.0 - 0.4} = 10.33$, it can be stated that $F\sin\theta = 10.3$.

3 a

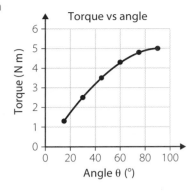

Torque vs angle

The curve of best fit looks like a sin curve, which is what would be anticipated from the relationship between the variables from the equation $\tau = rF\sin\theta$ when both F and r are kept constant.

b

Angle θ (°)	Torque (N m)	sin θ
15	1.3	0.259
30	2.5	0.500
45	3.5	0.707
60	4.3	0.866
75	4.8	0.966
90	5	1.000

c The gradient of the line of best fit represents $\dfrac{\text{rise}}{\text{run}}$ for the plotted data. This is equivalent to $\dfrac{\tau}{\sin\theta}$. From the equation, it can be seen that $\dfrac{\tau}{\sin\theta} = Fr$. Since the gradient is approximately $\dfrac{3.0 - 0}{0.6 - 0} = 5.0$, it can be stated that

$Fr = 5.0$.

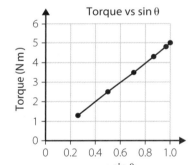

Torque vs sin θ

~continued in right column ▲

Chapter 3: Motion in gravitational fields

WS 3.1 PAGE 21

1 $a_c = \dfrac{v^2}{r}$

$$= \dfrac{\left(\dfrac{26\,528}{3.6}\right)^2}{(6370 + 1000) \times 10^3}$$

$$= 7.390\,01 = 7.390\,\text{m s}^{-2}$$

2 $F_c = \dfrac{mv^2}{r}$

For Mercury, $F_c = \dfrac{mv^2}{r} = \dfrac{3.30 \times 10^{23} \times (47.4 \times 10^3)^2}{57.9 \times 10^9}$

$$= 1.28 \times 10^{22}\,\text{N}$$

For Pluto, $F_c = \dfrac{mv^2}{r} = \dfrac{1.46 \times 10^{22} \times (4.7 \times 10^3)^2}{5910 \times 10^9}$

$$= 5.46 \times 10^{16}\,\text{N}$$

The centripetal force acting on Mercury is nearly 250 000 times greater than the centripetal force acting on Pluto.

3 $v = \dfrac{2\pi r}{T}$

$$= \dfrac{2\pi \times 9\,377\,000}{7.66 \times 3600}$$

$$= 2136.55 = 2140\,\text{m s}^{-1}$$

4 $v = \dfrac{2\pi r}{T}$, thus $\dfrac{107\,280}{3.6} = \dfrac{2\pi \times 149.6 \times 10^9}{T}$

So $T = \dfrac{2\pi \times 149.6 \times 10^9}{\left(\dfrac{107\,280}{3.6}\right)} = 31\,542\,433.6\,\text{s} = 365.1\,\text{days}$

5 $F = \dfrac{GMm}{r^2}$

$$= \dfrac{6.67 \times 10^{-11} \times 6.0 \times 10^{24} \times 75}{((1800 + 6370) \times 10^3)^2}$$

$$= 449.670 = 450\,\text{N}$$

6 $F_c = F_g$, so $\dfrac{m v^2}{\cancel{r}} = \dfrac{GM\cancel{m}}{r^{\cancel{2}}}$

Dividing both sides by m and multiplying both sides by r gives $v^2 = \dfrac{GM}{r}$.

Since M is the mass of Earth $= M_{\text{Earth}}$, $v = \sqrt{\dfrac{GM_{\text{Earth}}}{r}}$.

7 a

Satellite	m (kg)	r (m)	T (s)	v (m s⁻¹)	a_c (m s⁻²)	F_c (N)	F_g (N)
ISS	420 000	6.74×10^6	5529	7.66×10^3	8.71	2.46×10^{13}	2.46×10^{13}
Moon	7.35×10^{22}	3.78×10^8	2.31×10^6	1029	0.00280	2.058×10^{20}	2.058×10^{20}
Optus	1350	7.85×10^6	6908	7140	6.49	8767	8767
Optus D3	2420	6.87×10^6	5656	7632	8.48	20 521	20 521
Oscar	500	7.37×10^6	6284	7369	7.37	3684	3684
Sky Muster	875	7.14×10^6	5992	7487	7.85	6869	6869
Wresat	940	4.2164×10^7	85 986	3081	0.225	211.5	211.5

9780170449687

There may be small variations between your answer and the published one depending on application of significant figures and rounding in calculation.

b Satellites are objects that orbit another object, effectively in uniform circular motion. Consequently, they must be under the influence of a net centripetal force directed towards the centre of that circle. This force can only be provided by gravity. Hence F_c must equal F_g.

WS 3.2 PAGE 23

1 LEOs are used in satellite phone communication, GPS location, internet, observation, spying and collecting remote-sensing data. Their location close to Earth's surface results in very short lag time for signals and ability for close observation.

2 $v = \sqrt{\dfrac{GM_{Earth}}{r}}$

$= \sqrt{\dfrac{6.67 \times 10^{-11} \times 6.0 \times 10^{24}}{(540 + 6370) \times 10^3}}$

$= 7610.26 = 7.6 \times 10^3 \, \text{m s}^{-1}$

3 Use the unrounded answer from Question 2.

$v = \dfrac{2\pi r}{T}$, so $7610.26 = \dfrac{2\pi \times 6910 \times 10^3}{T}$

$T = \dfrac{2\pi \times 6910 \times 10^3}{7610.26} = 5705.036 = 5.7 \times 10^3 \, \text{s}$

4 $\dfrac{r^3}{T^2} = \dfrac{GM}{4\pi^2}$, so $\dfrac{r^3}{86\,164^2} = \dfrac{6.67 \times 10^{-11} \times 6.0 \times 10^{24}}{4\pi^2}$

$r = 4.222 \times 10^7$

Altitude $= r - r_{Earth}$
$= 4.222 \times 10^7 - 6.3781 \times 10^6$
$= 3.58419 \times 10^7 = 3.6 \times 10^7 \, \text{m}$

5 For a geosynchronous satellite,

$v = \dfrac{2\pi r}{T}$

$= \dfrac{2\pi \times 42\,220\,465}{86\,164}$

$= 3078.7 = 3.1 \times 10^3 \, \text{m s}^{-1}$.

The geosynchronous satellite moves at a speed significantly less than that of the Hubble Space Telescope.

6 Geosynchronous satellites are used for satellite and cable TV, weather forecasting, mapping, communication with spacecraft and as reference points for the GPS system. A satellite that is positioned at a great distance from Earth will have a 'footprint' over much more of the surface of Earth than one that is closer; also it will be much easier to locate reliably every day due to its orbital period (which is related to that distance).

7 Geostationary satellites maintain the same position, relative to the Earth observer, in the sky at all times during the day and night – they will appear stationary. Geosynchronous satellites move relative to the observer but will pass the same point in the sky at the same time each day.

8 The orbital period of the ISS is slightly over 90 minutes. That means it makes nearly 15 laps of Earth each day. It will tend

to be visible for several minutes once or twice per night to a viewer in eastern Australia, but would 'rise' and 'set' in different locations.

WS 3.3 PAGE 25

1 All planets move in elliptical orbits with the Sun as one focus.

2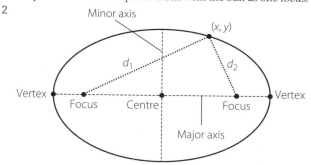

An ellipse is a closed curved shape enclosing two points called foci. It is the shape that results when the sum of the distances from the foci, $d_1 + d_2$, is kept constant.

3 Eccentricity is a measure of how far the shape of an ellipse departs from a circle; the eccentricity of a circle is 0. Venus has a very small eccentricity and so its orbit is very nearly circular. Since Earth has an eccentricity of nearly 0 we would expect its orbit to be approximately circular as well. The values for the other planets are somewhat higher and so their orbits would be less circular. Mercury has an eccentricity of greater than 0.2, by far the largest value for the planets listed, so its orbit would be much more elliptical and the distance between Mercury and the Sun would vary much more.

4 The total energy of a planet in orbit is constant. This total energy is the sum of kinetic and gravitational potential energies. As the planet gets closer to the Sun its gravitational potential energy decreases (becomes more negative) and, therefore, its kinetic energy must increase by the same amount. Hence it moves faster when closer to the Sun. This results in the planet moving through a greater angle as measured from the Sun but, as distance to the planet is so much less, the area is equivalent to the area swept in the same period by the planet when further away and moving more slowly.

5 Since G and $4\pi^2$ are constants, and M will be constant for all the planets in the same solar system, then $\dfrac{r^3}{T^2}$ must be a constant for all planets in the solar system. Put another way, r^3 is proportional to T^2 for all planets.

6 a Gradient of the line of best fit will be $\dfrac{r^3}{T^2} = \dfrac{GM}{4\pi^2}$ where M is the mass of the Sun.

Gradient $= \dfrac{(120 - 0) \times 10^{33}}{(350 - 0) \times 10^{13}} = 3.43 \times 10^{19}$

$M = \text{gradient} \times \dfrac{4\pi^2}{G} = 2.03 \times 10^{31} \, \text{kg}$

b This experimental value is out by a factor of 10. Alternatively, this could be expressed as a 90% error. This is a substantial error that means accuracy is poor. It would suggest that an error was made in data collection or manipulation.

1   There may be small variations between your answer and the published one depending on application of significant figures and rounding in calculation.

Planet	r (m) $\times 10^9$	T (days)	m (kg) $\times 10^{24}$	U (J) $\times 10^{32}$	E_{Total} (J) $\times 10^{32}$	v_{escape} (kms^{-1})
Mercury	57.9	88.0	0.330	−7.553	−3.777	4.3
Venus	108.2	224.7	4.87	−59.86	−29.93	10.4
Earth	149.6	365.2	5.97	−52.92	−26.46	11.2
Mars	227.9	687	0.642	−3.748	−1.874	5.0
Jupiter	778.6	4331	1898	−3249	−1625	59.5
Saturn	1433.5	10747	568	−528.6	−264.3	35.5
Uranus	2872.5	30581	86.8	−40.18	−20.09	21.3
Neptune	4495.1	59800	102	−30.30	−15.15	23.5

2   $F_c = F_g$ so $\dfrac{m v^2}{\cancel{r}} = \dfrac{GMm}{r^2}$, giving $v^2 = \dfrac{GM}{r}$

But $v = \dfrac{2\pi r}{T}$, so $v^2 = \dfrac{4\pi^2 r^2}{T^2} = \dfrac{GM}{r}$

Rearranging gives $\dfrac{r^3}{T^2} = \dfrac{GM}{4\pi^2}$, as required.

3   a   Gravitational force is given by the equation $F = \dfrac{GMm}{r^2}$.

If $r \to \infty$ then $F \to 0$ as anything divided by infinity is zero.

b   If the object is experiencing a gravitational force of zero then it is not in a gravitational field and thus has zero gravitational potential energy.

c   Work would need to be done to raise object m from a position close to object M to an infinite distance from M. If work has been done, then the gravitational potential energy of the object has increased. If the energy of m has increased from some value up to zero, then that value must have been negative in the first place.

Alternatively, negative work would have to be done to move a stationary object from an infinite distance from M to be stationary at a position closer to M (since gravity will do the work for us). If we have added negative energy to a mass that originally had zero energy, then it must now have negative energy. It is stationary, so its gravitational potential energy must be negative.

4   a   If $v = \sqrt{\dfrac{GM}{r}}$, then $v^2 = \dfrac{GM}{r}$.

Substituting in $K = \dfrac{1}{2} m v^2$, then $K = \dfrac{1}{2} m \times \dfrac{GM}{r} = \dfrac{GMm}{2r}$.

b $E_{total} = U + K = -\dfrac{GMm}{r} + \dfrac{GMm}{2r} = -\dfrac{GMm}{r}$

5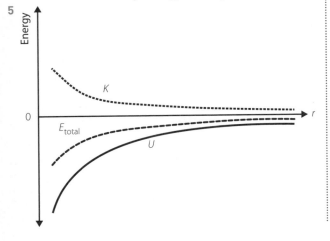

6   $E_{total} = U + K = -\dfrac{GMm}{r} + \dfrac{GMm}{2r} = -\dfrac{GMm}{r}$

At altitude 340 km,

$E_{total} = -\dfrac{6.67 \times 10^{-11} \times 420 \times 10^3 \times 6.0 \times 10^{24}}{2 \times (340 + 6370) \times 10^3}$

$= -1.2525 \times 10^{13}$ J

At altitude 420 km,

$E_{total} = -\dfrac{6.67 \times 10^{-11} \times 420 \times 10^3 \times 6.0 \times 10^{24}}{2 \times (420 + 6370) \times 10^3}$

$= -1.2377 \times 10^{13}$ J

Energy difference $= 1.48 \times 10^{11}$ J

7   a   The minimum initial speed must be enough to just reach a distance of infinity – in other words, to reach there and stop so that $K = 0$. At an infinite distance there is no gravitational force acting on the object, so it is not in a gravitational field and $U = 0$.

b   Given that $U_{final} = 0$ and $K_{final} = 0$, then $E_{total\ final} = 0$
Thus $E_{total\ initial} = 0$ and $U_{initial} + K_{initial} = 0$

Therefore $-\dfrac{GMm}{r} + \dfrac{m v^2}{2} = 0$

$\dfrac{v^2}{2} = \dfrac{GM}{r}$

$v^2 = \dfrac{2GM}{r}$

$v = \sqrt{\dfrac{2GM}{r}}$

MODULE FIVE: CHECKING UNDERSTANDING PAGE 30

1   **D**   The centripetal force provided by the sideways frictional force from the tyres must be greater on Bend A than on Bend B.

2

	Orbital velocity	Orbital period	Centripetal acceleration
D	A is greater	B is greater	A is greater

3   **A**   all planets in the solar system.

4 $s_x = u_x t$

$135 = u\cos 30° \times t$

$t = \dfrac{135}{u\cos 30°}$ and

$s_y = u_y t + \dfrac{1}{2}at^2$

$0 = u\sin 30° \, t - 4.9t^2$

Thus $u\sin 30° = 4.9t$

Substituting for t:

$u\sin 30° = 4.9 \times \dfrac{135}{u\cos 30°}$

$u^2 = \dfrac{4.9 \times 135}{\sin 30° \cos 30°}$

$u = 39.085 = 39\,\text{m s}^{-1}$

5 The centripetal force is equivalent to the horizontal component of the normal force (N_x). The vertical component of the normal force (N_y) is equivalent to the weight force (mg).

Thus $N_y = mg = N\sin 15°$ and $N_x = N\cos 15° = F_c = 30\,400\,\text{N}$

Thus $\tan 15° = \dfrac{mg}{30\,400}$

Therefore $m = 831.189 = 830\,\text{kg}$

6 a Using $v = \dfrac{2\pi r}{T}$, $r = \dfrac{Tv}{2\pi} = 5023\,\text{m}$

Since $a_c = \dfrac{v^2}{r}$ and $a_c = g$, then

$g = \dfrac{v^2}{r} = \dfrac{8.167^2}{5023}$

$= 0.013\,278\,89 = 0.013\,28\,\text{m s}^{-2}$

b Also, $g = \dfrac{GM}{r^2}$

so $M = \dfrac{gr^2}{G} = \dfrac{0.013\,28 \times 5023^2}{6.67 \times 10^{-11}}$

$= 5.022\,99 \times 10^{15} = 5.02 \times 10^{15}\,\text{kg}$

MODULE SIX: ELECTROMAGNETISM

REVIEWING PRIOR KNOWLEDGE PAGE 32

1 a

b Possible answers include:

- They originate on positive charges and terminate on negative charges.
- They cannot cross.
- The density of lines is proportional to the electric field strength. For example, in our answer to **a**, the fields lines are equally spaced everywhere within the plates, so the field strength is constant.
- They are perpendicular to equipotential lines – equipotentials are lines or planes of constant electric potential.
- Positive charges will experience a force in the direction of the field lines.
- Negative changes will experience a force in the opposite direction to the field lines.

This list is not exhaustive. Other answers may be acceptable.

c Use $E = -\dfrac{\Delta V}{d}$ where d is the plate separation and ΔV is the potential difference from one plate to the other.

$E = -\dfrac{175\,\text{V}}{2.5 \times 10^{-2}\,\text{m}}$

$= -7000\,\text{V m}^{-1} = -7.0\,\text{kV m}^{-1}$

The magnitude is $7.0\,\text{kV m}^{-1}$ and the direction is from the positive to the negative plate.

d If we recall the definition if the electron volt, we can write down our answer immediately – if an electron is accelerated through a potential difference of 1 V, it gains 1 eV of energy. Our potential difference is 175 V, so our electron gains 175 eV.

For joule, we multiply by the charge of an electron:

$E = (175\,\text{V})(1.6 \times 10^{-19}\,\text{C}) = 2.8 \times 10^{-17}\,\text{J}$

We could also use $\Delta U = q\Delta V$. Note that the charge is negative but so is the change in potential, because potential increases in the opposite direction to the direction of the field lines. That is, potential is lower at the positive plate for the electron, which is where the electron ends up, and $\Delta V = V_{\text{final}} - V_{\text{initial}}$.

e The energy gained by the electron is kinetic energy. Therefore, because the change in kinetic energy is equal to the work done on the electron, we can write:

$\dfrac{1}{2}mv^2 = q\Delta V$

Rearranging to make v the subject:

$v = \sqrt{\dfrac{2q\Delta V}{m}} = \sqrt{\dfrac{2(-175\,\text{V})(-1.6 \times 10^{-19}\,\text{C})}{9.11 \times 10^{-31}\,\text{kg}}} = 7.8 \times 10^6\,\text{m s}^{-1}$

f $\vec{F} = q\vec{E}$

g We know $u = 0\,\text{m s}^{-1}$ and $v = 7.8 \times 10^6\,\text{m s}^{-1}$. We also know s, the spacing (2.5 cm). We want to find t. The equations $v = u + at$ and $v^2 = u^2 + 2as$ link these quantities. $u = 0$ so $v = at$, thus $a = \dfrac{v}{t}$. Then:

$v^2 = u^2 + 2as = 2\left(\dfrac{v}{t}\right)s$

We then make t the subject:

$t = \dfrac{2s}{v} = \dfrac{0.05\,\text{m}}{7.840\,34 \times 10^6\,\text{m s}^{-1}} = 6.4 \times 10^{-9}\,\text{s}$

2 a We relate the current in a conductor (I) to the potential difference across the conductor (V) and the electrical resistance of the conductor (R) using Ohm's Law, $V = IR$.

b We rearrange the equation to get:

$I = \dfrac{V}{R} = \dfrac{232\,\text{V}}{25\,\Omega} = 9.3\,\text{A}$

Note the result is given to two significant figures because the least precise input value has two figures.

c The power dissipated is given by

$P = IV = I^2 R = \dfrac{V^2}{R}$

$= \dfrac{(232\,\text{V})^2}{25\,\Omega} = 2153\,\text{W} = 2.2\,\text{kW}$

3 a Imagine gripping the wire in your right fist, with your thumb pointing along the current direction. The magnetic field lines will be concentric circles, with the field lines pointing in the direction of your curled fingers.

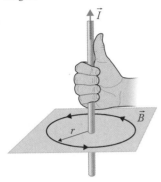

b Imagine gripping a horizontal loop of wire in your right fist, with your thumb pointing along the current. Your fingers will either be pointing down inside the loop and up outside or vice-versa. The field lines follow your fingers.

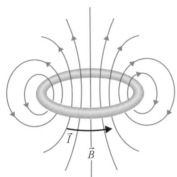

Chapter 4: Charged particles, conductors and electric and magnetic fields

WS 4.1 PAGE 35

1 a Positive charge experiences a force that follows the field – down.

b Negative charge experiences a force up.

c Neutron is uncharged and does not experience a force.

2 a Recall that the units of electric field can be expressed as Vm^{-1} or NC^{-1}.
$$F = qE$$
$$= (1.602 \times 10^{-19}\,C)(100\,N\,C^{-1})$$
$$= 1.60 \times 10^{-17}\,N$$

b We are interested in the magnitude of the force. Because an electron has the same magnitude of charge as the proton, the magnitude of the force will be the same: $1.60 \times 10^{-17}\,N$.

c No charge so no force. $0\,N$

3 a Newton's second law: $F = ma = qE$
$$a = \frac{qE}{m}$$
$$= \frac{(1.602 \times 10^{-19}\,C)(100\,NC^{-1})}{1.673 \times 10^{-27}\,kg}$$
$$= 9.58 \times 10^{9}\,m\,s^{-2} \text{ (to 3 significant figures because 100 has 3 signficant figures)}$$

b As for a, but with the electron mass:
$$a = \frac{qE}{m}$$
$$= \frac{(1.602 \times 10^{-19}\,C)(100\,NC^{-1})}{9.109 \times 10^{-31}\,kg}$$
$$= 1.76 \times 10^{13}\,m\,s^{-2}$$

c No charge so no force and no acceleration. $0\,m\,s^{-2}$

4 a Substitute the numbers, recalling that initial kinetic energy is zero:
$$\Delta K = K_{final} - K_{initial} = qEd$$
But $K_{initial} = 0$, so
$$K_{final} = qEd$$
$$= (1.602 \times 10^{-19}\,C)(100\,N\,C^{-1})(0.001\,00\,m)$$
$$= 1.60 \times 10^{-20}\,J$$
Again, we may consider the units: the coulombs cancel to leave $N\,m$, which is joule.

b The mass does not enter into it, so the electron will have the same kinetic energy as the proton. It is much lighter, though, so it will be going much faster.

5 The neutron has no charge, does not experience a force and so does not move 1 mm, so the calculation cannot be done. We know it has zero kinetic energy, but $d = 0$.

6 a The field is vertically downwards. That means the higher potential is above, the lower below. The field is measured in Vm^{-1}, so for every 1 m we increase the potential by 100 V. Therefore, if we rise 2.5 m, the potential is 250 V higher. We started at 250 V, so the total is 250 V + 250 V = 500 V.

b At 2.5 m lower the potential has fallen by 250 V, so it is now zero.

c We do not change height when moving to the right. Because it is a uniform field, the potential must be unchanged; 250 V.

d The movement right does not matter, only the change in height; 3.5 m gives an extra 350 V, or a potential of 600 V.

7 The proton gains kinetic energy. Therefore, positive work was done on it by the electric field.

8 The electron gains kinetic energy. Therefore, positive work was done on it by the electric field.

WS 4.2 PAGE 38

1 a No. There is no horizontal force and therefore, because $F = ma$, $a = 0$.

b Yes, the electron and proton will both accelerate vertically because both will experience a vertical force. The proton will follow the field and therefore accelerate down. The electron accelerates up.

2 In all cases the horizontal component is constant at $112\,m\,s^{-1}$. We must use constant acceleration equations for the vertical components, although we can see the neutron will not accelerate, being neutral.

a Consider only vertical (y) components.
$$v_y = u_y + a_y t$$
$$= u_y + \frac{qEt}{m}$$
Because $a = \frac{qEt}{m}$ and $u_y = 0$, we can write
$$v_y = \frac{(1.602 \times 10^{-19}\,C)(100\,N\,C^{-1})(1.00 \times 10^{-9}\,s)}{1.675 \times 10^{-27}\,kg}$$
$$= 9.56\,m\,s^{-1} \text{ (down) to 3 significant figures because 1.00 ns has 3 figures.}$$

b We do the same calculation as in a, but with a different mass:
$$v_y = \frac{(1.602 \times 10^{-19}\,C)(100\,N\,C^{-1})(1.00 \times 10^{-9}\,s)}{9.109 \times 10^{-31}\,kg}$$
$$= 1.76 \times 10^{4}\,m\,s^{-1} \text{ (up)}$$

c Zero, as noted.

3 a The proton has a velocity $112\,\text{m s}^{-1}$ to the right (x) and $9.58\,\text{m s}^{-1}$ down (y). The total magnitude can be found from Pythagoras's theorem:

$$v = \sqrt{(v_x)^2 + (v_y)^2}$$

$$= \sqrt{(112\,\text{m s}^{-1})^2 + (9.58\,\text{m s}^{-1})^2}$$

$$= 112\,\text{m s}^{-1}$$

It can be seen that the vertical component of velocity is small compared to the horizontal component of velocity and does not affect the total (to 3 significant figures). The direction is affected.

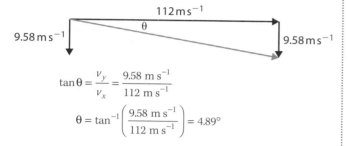

$$\tan\theta = \frac{v_y}{v_x} = \frac{9.58\,\text{m s}^{-1}}{112\,\text{m s}^{-1}}$$

$$\theta = \tan^{-1}\left(\frac{9.58\,\text{m s}^{-1}}{112\,\text{m s}^{-1}}\right) = 4.89°$$

So, after $1.00\,\text{ns}$, the proton is heading to the right, about 5° below the horizontal. This is an important example – even though one component has little effect on the overall *speed*, it still affects the *direction*.

b The electron has a velocity $112\,\text{m s}^{-1}$ to the right and $1.76 \times 10^4\,\text{m s}^{-1}$ up. The horizontal component is negligible compared to the vertical, so we can effectively say its velocity is $1.76 \times 10^4\,\text{m s}^{-1}$ up. The angle is also very small – about 0.4° right of vertically up, as can be seen from the working below.

$$v = \sqrt{(v_x)^2 + (v_y)^2}$$

$$= \sqrt{(1.76 \times 10^4)^2 + (112)^2}$$

$$= 1.76 \times 10^4\,\text{m s}^{-1}$$

$$\tan\theta = \frac{v_y}{v_x} = \frac{1.76 \times 10^4}{112}$$

$$\theta = \tan^{-1}\left(\frac{1.76 \times 10^4}{112}\right) = 89.6°$$

c For the neutron, the velocity remains $112\,\text{m s}^{-1}$ to the right.

4 We use $s = ut + \frac{1}{2}at^2$ to show that the position versus time will be parabolic. We know that the vertical velocity is initially zero so, using Newton's second law, the parabola can be described by the equation $s_y = \frac{qEt^2}{2m}$, where $s_y = 0$ when $t = 0$. The field is down, so we make sure our graph makes it clear which way is up!

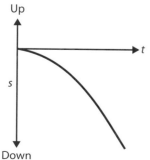

When dealing with such small particles accelerating in electric fields, the speeds can get very high. When particle velocities become an appreciable fraction of the speed of light, relativistic effects must be taken into account. Refer to Chapter 12 for a discussion of special relativity. We do not tackle relativistic examples in Module 6.

- - -

WS 4.3 PAGE 40

1 a We use $F = qvB\sin\theta$, but the angle between the velocity and the field, θ, is zero, and $\sin 0° = 0$, so there is no force.

b We use $F = qvB\sin\theta$ but now $\theta = 90°$, so $\sin\theta = 1$ and $F = qvB$. The magnitude of the force is then:

$$F = qvB$$

$$= (1.602 \times 10^{-19}\,\text{C})(1100\,\text{m s}^{-1})(4.00\,\text{T})$$

$$= 7.05 \times 10^{-16}\,\text{N}$$

2 In all cases, the force is obtained from $F = qvB\sin\theta$. The acceleration is given by $a = \frac{F}{m}$, so that $a = \frac{q}{m}vB\sin\theta$, where we take the charge-to-mass ratio to the front. The remainder of the expression involving v, B and the angle is the same for all particles, so can be ignored.

For the neutron, $q = 0$, so we can ignore it. Examining the $\frac{q}{m}$ ratio for the other three particles it can seen that the proton and electron have equal charge whereas the alpha particle has twice the charge. The alpha particle is four times the mass of a proton and the electron has a mass approximately 1760 times smaller. Consequently, the $\frac{q}{m}$ ratio of the electron is very large compared to that of the proton, while the alpha has a $\frac{q}{m}$ ratio half that of the proton. This means the electron experiences the greatest acceleration.

3 a We know that when the velocity is perpendicular to the magnetic field, the resulting force is perpendicular to both. That means the resulting acceleration is perpendicular to both. When we have constant velocity perpendicular to a constant acceleration, we get circular motion (see Chapter 3).

Circular motion obeys the relation $a_c = \frac{v^2}{r}$ and $a_c = \frac{F}{m}$, so that $\frac{F}{m} = \frac{v^2}{r}$ and so $r = \frac{mv^2}{F}$. This expression makes sense – if the particle is moving faster or is more massive, it will be harder to turn and the radius will be greater. If the force is large, the radius will be smaller.

Now, we just substitute in the expression for F due to a perpendicular magnetic field to get

$$r = \frac{mv^2}{qvB}$$

$$= \frac{mv}{qB} = \left(\frac{v}{B}\right)\left(\frac{q}{m}\right)^{-1}$$

So we see that the radius is inversely proportional to the charge-to-mass ratio, $\frac{q}{m}$.

b If we create a beam of charged particles of uniform velocity then pass it into a region of perpendicular magnetic field, the particles will have curved paths. The radius of the path for each particle will depend on the charge-to-mass ratio of the particle. Particles of opposite sign will curve in opposite directions. This means the different particles (say protons, alpha particles, deuterons and so on) will follow different paths, and we can count the numbers of each using separate detectors.

c An alpha particle is a nucleus of ^{4}He, so it contains two protons and two neutrons and has a charge equal to that of two protons. A deuteron is a nucleus of ^{2}H, and contains one proton and one neutron, and has a charge from one proton. It has very close to half the mass of the alpha particle, but half the charge as well, so these two particles would be difficult to separate using a mass spectrometer because the $\dfrac{q}{m}$ ratio would be very similar.

Chapter 16 explains why the alpha particle does not have *exactly* double the mass of the deuteron.

4 We can think about our results from mechanics. $W = Fs\cos\theta$, where θ is the angle between the force, F, and the displacement, s. The magnetic field induces circular motion. In circular motion, the force is perpendicular to the velocity. The displacement, s, is perpendicular to F, so $\theta = 90°$ and $\cos\theta = 0$.

5 The particle has a velocity component parallel with the field and another velocity component perpendicular to the field. The parallel component gives rise to no force. The perpendicular component gives rise to a force that results in circular motion whose radius depends on the perpendicular velocity, the charge-to-mass-ratio of the particle and the magnetic field strength. There is no force parallel to the field, so the particle velocity parallel to the field remains constant. The resulting motion is a helix (something like a coiled spring) of constant pitch and radius (see diagram), and whose axis is parallel to the field direction. See Figure 5.17 in the *Physics in Focus: Year 12* student book.

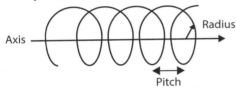

WS 4.4 **PAGE 42**

1 a We have two forces, $F_g = mg$ (down) and $F_E = qE$. If the particle does not accelerate, the net force is zero, so we can say:

$$F_g + F_E = mg + qE = 0$$

or

$$E = \frac{-mg}{q} = \frac{-(0.23 \times 10^{-3} \times 10^{-3}\ \text{kg})(9.8\ \text{m s}^{-2})}{-11.3 \times 10^{-9}\ \text{C}}$$

$$= 2.0 \times 10^2\ \text{V m}^{-1}$$

Remembering to convert to kg and C, we have taken down as positive here (by taking g as positive), so the required electric field points down. This makes sense because we have a negative charge, so the force on it due to the field will be up, counteracting gravity.

b We use the same idea but with the magnetic force, where $\sin\theta = 1$.

$$F_g + F_B = mg + qvB = 0$$

$$B = \frac{-mg}{qv} = \frac{-(0.23 \times 10^{-3} \times 10^{-3}\ \text{kg})(9.8\ \text{m s}^{-2})}{(-11.3 \times 10^{-9}\ \text{C})(1000\ \text{m s}^{-1})} = 0.2\ \text{T}$$

The force on the particle must be upwards. We can use the right-hand rule to show that the field direction must be horizontal and perpendicular to the velocity. Because the charge is negative, the field is right-to-left relative to the direction of the moving charge (see Figure 5.15 in the *Physics in Focus: Year 12* student book).

2 $F_g + F_E = mg + qE = 0$ and $E = -\dfrac{\Delta V}{d}$

$$mg - \frac{q\Delta V}{d} = 0$$

$$q = \frac{mgd}{\Delta V}$$

$$= \frac{(2.7 \times 10^{-3}\ \text{kg})(9.8\ \text{m s}^{-2})(0.50\ \text{m})}{20 \times 10^3\ \text{V}} = 6.6 \times 10^{-7}\ \text{C}$$

3 A cyclotron is a particle accelerator.

A cyclotron body consists of electrodes, called 'dees' because of their shape. These dees sit in a narrow gap between poles of a large magnet, which creates a perpendicular magnetic field. A stream of charged particles is fed into the centre of the chamber and a high frequency alternating voltage is applied across the electrodes. This voltage creates an electric field that alternately attracts and repels the charged particles, causing them to accelerate across the gap between the oppositely charged dees.

The magnetic field moves the particles in a circular path and, as they gain more energy from the accelerating voltage, they spiral outwards until they reach the outer edge of the chamber.

4 The particles will travel in a straight line if the magnetic and electric fields exert equal and opposite forces on the charged particles. For this to occur, the magnetic field will have to be perpendicular to both the electric field and the motion of the charged particles. This means that $qE = qvB$ (assuming the particles move at right angles to the magnetic field) and, therefore, $v = \dfrac{E}{B}$. Do your answers to questions **1a** and **1b** agree with this?

Chapter 5: The motor effect

WS 5.1 **PAGE 44**

1 Use the expression for the force, $F = lIB\sin\theta$. When the wire is parallel, $\theta = 0$ and $\sin\theta = 0$. When the wire is perpendicular, $\theta = 90°$ and $\sin\theta = 1$, so F reaches its maximum value of lIB. Hence a sketch of a sine curve of this amplitude for $0 \le \theta \le 90°$ is appropriate.

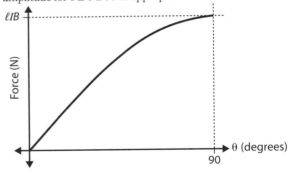

2 a No current means no magnetic force.

b Current is parallel with the field. Still no magnetic force.

c The magnitude of the force is $F = lIB\sin\theta$, and now $\theta = 90°$ so:

$$F = lIB = (0.250\ \text{A})(0.0100\ \text{m})(0.0550\ \text{T})$$

$$= 1.375 \times 10^{-4}\ \text{N} = 1.38 \times 10^{-4}\ \text{N}$$

For the direction, we apply the right-hand rule: thumb pointing in the direction of the current (out of page), the fingers along the field (left to right). The palm gives the direction of the force – up the page.

d The magnitude is the same as in **c** because all the input values, including the angle, are the same. We apply the right-hand rule: thumb down the page, fingers along the field (left to right). The palm gives the direction of the force – out of the page.

WS 5.2 **PAGE 46**

1 The field is given by $B = \dfrac{\mu_0 I}{2\pi r}$ where r is the distance from the centre of the wire. So, the form of the curve is $\dfrac{1}{r}$. We are told not to plot for $r \le R$, and at R the field magnitude will be $B = \dfrac{\mu_0 I}{2\pi R}$.

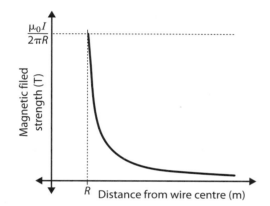

2 We can multiply force per unit length by the length of the wire to get the total force. Force per unit length is independent of the actual length of the wire – it does not vary with the length of the wire and depends only on the current. Because the mass of the wire also increases with length, we can use mass per unit length and force per unit length to get the acceleration of the wire regardless of its length.

3 Force per unit length is given by

$$\frac{F_{\text{by 1 on 2}}}{\ell} = \frac{\mu_0 I_1 I_2}{2\pi r}$$
$$= \frac{(1.256\,637\,06\times10^{-6})(1.0)(2.5)}{2\pi(0.50)}$$
$$= 1.0\times10^{-7}\,\text{N m}^{-1}$$

The currents are in opposite directions, so the force is repulsive.

4 Yes. The ability of the wires to respond to the force does not affect the magnitudes of the forces.

5 To factor out the effects of the ends of the wires – where the current direction must change – the definition imagines wires of infinite length. We cannot calculate the force on an infinite wire, but we can calculate the force per metre on such a wire.

Chapter 6: Electromagnetic induction

WS 6.1 **PAGE 48**

1 There are three variables on the right side of the equation: B, A and θ. So we can change Φ by changing the area of the loop, the magnitude of the magnetic field, or the angle between the field and the normal to the loop.

2 a The equation for flux is $\Phi = BA\cos\theta$, so the curve is a cosine curve. θ in the equation is the angle between the field lines and the normal to the plane of the area,

which is the same definition as used in the question. At the origin, $\theta = 0°$, or $\cos\theta = 1$, the maximum, where the magnitude will be BA.

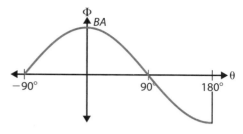

The graph has its maximum value at $\theta = 0°$ and $\theta = 180°$ and goes to zero at $\theta = \pm90°$.

b When the field is parallel to the normal, it is perpendicular to the loop and the flux will be a maximum.
$$\Phi = BA\cos\theta$$
$$= (1.0\,\text{T})(0.25\times0.25\,\text{m}^2)\cos0°$$
$$= 0.0625 = 0.063\,\text{Wb}$$

c $\Phi = BA\cos\theta$
$$= (1.0\,\text{T})(0.25\times0.25\,\text{m}^2)\cos34°$$
$$= 0.052\,\text{Wb}$$

d If the field is perpendicular to the normal it is parallel to the loop, so no field lines pass through the loop and the flux is $0\,\text{Wb}$.

3 The field and the angle are constant, so the rate of change of flux comes from the rate of change of area. So:

$$\frac{\Delta\Phi}{\Delta t} = B\cos0°\left(\frac{\Delta A}{\Delta t}\right)$$

The rate of change of the area is the area swept out by the bar in one second.

The bar will move distance s in time t, where $s = vt$. Multiplying this distance by the length of the bar, ℓ, yields the area swept out:

$A = vt\ell$

To determine the rate of change of area, divide this by the time.

$$\frac{\Delta A}{\Delta t} = \frac{vt\ell}{t} = v\ell$$

Note: $5.5\,\text{cm} = 0.055\,\text{m}$. The diagram shows the area is decreasing, so the velocity is given a negative sign.

$$\frac{\Delta A}{\Delta t} = (0.055\,\text{m})(-0.50\,\text{m s}^{-1})$$
$$\frac{\Delta\Phi}{\Delta t} = (0.25\,\text{T})(0.055\,\text{m})(-0.50\,\text{m s}^{-1})$$
$$= -0.0069\,\text{Wb s}^{-1}$$

WS 6.2 **PAGE 50**

1 We use $\varepsilon = -\dfrac{\Delta\Phi}{\Delta t} = -\dfrac{\Phi_{\text{final}} - \Phi_{\text{initial}}}{t_{\text{final}} - t_{\text{initial}}}$ and apply it to each segment.

a $\varepsilon = -\dfrac{\Delta\Phi}{\Delta t} = -\dfrac{\Phi_{\text{final}} - \Phi_{\text{initial}}}{t_{\text{final}} - t_{\text{initial}}}$
$$= -\frac{(14-2)\,\text{Wb}}{(5-0)\,\text{s}} = -2.4\,\text{V}$$

b $\varepsilon = -\dfrac{(14-14)\,\text{Wb}}{(11-5)\,\text{s}} = 0\,\text{V}$

c $\varepsilon = -\dfrac{(2-14)\,\text{Wb}}{(21-11)\,\text{s}} = 1.2\,\text{V}$

d There are three regions, each with a constant slope. So, the graph consists of three lines each of constant value (that is, horizontal).

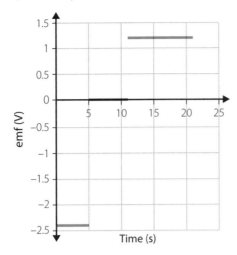

e If the loop is replaced with a coil, the magnetic flux is passing through 15 loops instead of 1 and, thus, the area is effectively 15 times as great. That will increase the magnitudes of the induced emfs. So in the first segment $-2.4 \times 15 = 36\,V$, in the second $0\,V$ and in the third $18\,V$.

2 a We use $\varepsilon = -\dfrac{\Delta\Phi}{\Delta t} = \dfrac{-\Delta(BA\cos\theta)}{\Delta t}$.

$\dfrac{\Delta\Phi}{\Delta t} = -0.0069\,\text{Wb s}^{-1}$, so

$$\varepsilon = -\dfrac{\Delta\Phi}{\Delta t}$$
$$= -(-0.0069\,\text{Wb s}^{-1}) = 0.0069\,\text{Wb s}^{-1}$$
$$= 0.0069\,\text{V}.$$

b As the bar slides to the left, the amount of flux through the loop into the page decreases. By Lenz's Law, the induced current will flow to increase the flux into the page, opposing the change.

Pointing the right thumb into the page, the fingers curl in the clockwise direction, showing that the current must increase in that direction.

WS 6.3 PAGE 52

1 Possible answers include:
- Step-down transformer (found on power poles, for example) required for stepping down the transmitted voltage to the 230 V AC for use in residences
- Step-down transformer required for reducing the 230 V to something suitable for charging phones, computers, individual batteries and so on (these include extra components to convert the AC house power to DC)
- Step-down transformer required for stepping down the 230 V house power to lower voltages (typically 12 V) used in some internal houselights, especially halogen downlights
- Step-down transformer required for stepping down the 230 V house power to lower voltages for silicon chips, which usually operate at low voltages and with DC current, used in any home appliance that includes electronics (such as a stereo system, TV, desktop computer, printer, washing machine – nearly everything)
- Step-up transformers for microwave ovens because they need *higher* voltages than the house supply, so use a transformer to increase the voltage

Appliancs and toys that are battery-powered usually do not use transformers, except for charging. Many appliances that use AC power from a wall socket do use a transformer.

2 a **i** For AC current, such a transformer can be used to give a secondary voltage higher than the primary voltage: T

  ii For constant DC current, such a transformer can be used to give a secondary voltage higher than the primary voltage: F

  iii When the secondary voltage is lower than the primary voltage, the secondary current is also lower than the primary current: F

  iv When the primary coil has more turns than the secondary coil, the primary coil has a higher current than the secondary coil: F

  v A step-up transformer has more turns in the secondary coil than in the primary coil: T

2 b **i** It is known that $\dfrac{V_P}{V_S} = \dfrac{N_P}{N_S}$, so for the secondary voltage to be higher, all we need is the secondary coil to have more turns than the primary.

  ii When the current is DC, there is no change in the magnetic flux due to the primary coil, because it has a constant current in it. That means $\dfrac{\Delta\Phi}{\Delta t} = 0$ and the transformer will not work.

  iii If the secondary voltage is lower, conservation of energy tells us that the secondary current will be higher.

  iv From our answer to i, $\dfrac{V_P}{V_S} = \dfrac{N_P}{N_S}$, so we know that if the primary has more turns, it will have higher voltage. If its voltage is higher, its current must be lower.

  v This comes from $\dfrac{V_P}{V_S} = \dfrac{N_P}{N_S}$ and the definition of a step-up transformer – a transformer than has a higher output than input voltage.

3 a First, we must find the current in the secondary coil. Combining $P = IV$ and $V = IR$ gives $P = I^2R$, or:

$$I = \sqrt{\dfrac{P}{R}} = \sqrt{\dfrac{1400\,\text{W}}{8.6\,\Omega}} = 12.7589\,\text{A},$$

Now, combine $\dfrac{V_P}{V_S} = \dfrac{N_P}{N_S}$ and $I_P V_P = I_S V_S$ to get:

$$N_P = N_S\left(\dfrac{I_S}{I_P}\right) = \dfrac{500 \times 12.7589\,\text{A}}{6.1\,\text{A}} = 1046\ \text{turns}$$

Given the precision of the data used, we might round this to 1050 turns.

b For the secondary coil, we have $P = IV$ where $P = 1400\,\text{W}$ and $I = 12.7589\,\text{A}$, giving

$$V_S = \dfrac{P_S}{I_S} = \dfrac{1400\,\text{W}}{12.7589\,\text{A}} = 109.7273\,\text{V}$$

$\dfrac{V_P}{V_S} = \dfrac{N_P}{N_S}$, which gives

$$V_P = V_S\left(\dfrac{N_P}{N_S}\right)$$

$$= 109.7273\left(\dfrac{1046}{500}\right)$$

$$= 229.55\,\text{V} = 230\,\text{V}$$

c It is a step-down transformer, because the output voltage is smaller than the input voltage.

d If the bar heater is designed to use 110 V power, it was probably designed for use in the USA. Because the voltage is being stepped down from 230 V to 110 V, it is probably being used in Australia.

Chapter 7: Applications of the motor effect

WS 7.1 PAGE 55

1 We use the right-hand rule. Fingers along the field, thumb along the current, force is out of the palm. So, we find:

   a down the page
   b up the page
   c up the page
   d up the page

   The diagram, with arrows added, should look something like this:

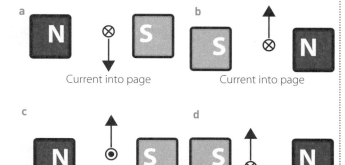

a Current into page

b Current into page

c Current out of page

d Current into page

2 a Use $\tau = nIAB\sin\theta$, where $n = 1$ and $\sin\theta = 1$, so
      $\tau = nIAB\sin\theta = IAB$
      $= (7.5\,\text{A})(2.00 \times 10^{-4}\,\text{m}^2)(0.500\,\text{T}) = 7.5 \times 10^{-4}\,\text{N}\,\text{m}$

   b Now $n = 128$, so:
      $\tau_{n=128} = 128 \times \tau_{n=1} = 128 \times 7.5 \times 10^{-4}\,\text{N}\,\text{m} = 9.6 \times 10^{-2}\,\text{N}\,\text{m}$

   c Because $\tau = nIAB\sin\theta$ and all but θ are constant, it can be seen that the maximum occurs when $\sin\theta = 1$; that is, $\theta = 90°$. It will be 15% of its maximum when $\sin\theta$ is 15% of its maximum – that is, 0.15.
      $\sin\theta = 0.15$, so $\theta = \sin^{-1} 0.15 = 8.6°$

3 More coils will result in more torque, which means a more powerful motor (albeit a heavier one).

   Another reason is best seen by reference to Figure 8.6 in the *Physics in Focus Year 12 Student Book*. If we have a single coil, the torque drops to almost zero before the commutator switches the current. The result is a very uneven delivery of torque and, specifically, certain angles at which the motor lacks torque. Coils offset from one another can result in more even torque throughout rotation.

   Other effects can also result. When the normal to the coil is parallel to the field, the torque is zero ($\tau = nIAB\sin\theta$ and $\theta = 0$). If the motor stops in this position, it cannot start itself spinning; it needs a push. And while it is not spinning, the current is still flowing but not doing any work, and neither is it inducing a back emf, so the motor might burn out from high current flow. When the motor is spinning, the back emf limits the magnitude of the current, preventing this.

4 a
commutator	B
permanent magnets	D
carbon brushes	F
coils of wire	E
armature	C
casing	A

   b armature, coils, commutator
   c casing, permanent magnets, brushes

WS 7.2 PAGE 58

1 The back emf in the motor is exactly the same in origin as the induced emf of a generator. In the motor, the coils are spun by the motor effect resulting from the supply emf, and the back emf works against the supply emf. In a generator, the coils are spun by an external driving force, such as a turbine or pedals, and there is no supply emf.

2 a The period, T, is given by $T = \dfrac{1}{f} = \dfrac{1}{60} = 0.017\,\text{s}$.

   b We have a cosine curve of period 0.017 s and maximum height value of 700 V. Since it starts with the normal to the plane of the coil perpendicular to the field, the emf is a maximum at $t = 0\,\text{s}$.

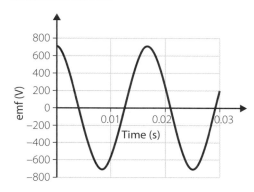

   c i Fewer turns would decrease the amplitude of the curve in proportion, because each turn of wire will generate its own emf as it experiences a change in flux. The total emf will be the sum of the emfs from each turn.

      ii Greater area would increase the amplitude of the curve in proportion. A larger area will experience a proportionally greater change in flux, resulting in an increased emf. (We can think of increasing the number of turns as an increase in area, so we expect area and turns to have similar effects.)

      iii Increasing the magnetic field would increase the amplitude of the curve in proportion. As magnetic field is increased the change in flux is proportionally increased, resulting in an increased emf.

      iv Faster spinning would increase the frequency. This would increase the rate at which the change in flux is experienced and thus increase the amplitude of the curve. There would also be a decrease in the period – that is, the maxima (high points) would get closer together on the graph.

d The negative parts of the graph would be mirrored upwards into the positive. The graph would look something like this:

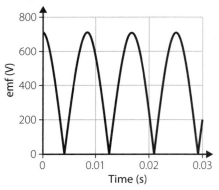

1 a Current will be highest when there is no back emf. This happens when the motor is not turning. This will be the case (1) immediately after it is turned on and (2) if the load is so great the motor stops turning.

b The current will be lowest when the back emf is greatest. This is when the motor is spinning freely at its top speed.

c In all cases, the current is given by:

$$\text{current} = \frac{\text{supply emf} - \text{back emf}}{\text{resistance}}.$$

Before the motor starts turning, back emf is zero and so, when $t = 0$, $\text{current} = \dfrac{\text{supply emf}}{R}$, which is its largest possible value.

In the next time period, the motor is accelerating up to its maximum speed, so the back emf is increasing from zero to its maximum value. This means I is also changing as a function of time.

We know that when the motor reaches its full speed back emf will be constant and at its largest. Therefore, the net emf will be at its lowest. So, for large t (when the motor has reached its top speed), the current will be at its lowest and constant.

As the motor spins faster, the back emf gets bigger, the current gets smaller, the torque on the motor gets smaller, the motor experiences a weaker rotational acceleration so it speeds up more slowly, and so back emf increases more slowly. Consequently, the curve will start steep and then flatten out gently.

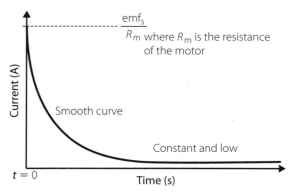

2 a If you shake the torch, the magnet passes through the coil. If the switch is closed, a current can flow in the coil. By Lenz's and Faraday's Laws, the magnetic flux in the coil is changing and so a current will flow to oppose the change. This current will pass through the bulb and generate light.

b If the switch is open, no current can flow, so no opposing force will act on the magnet, the globe cannot glow and the magnet will move freely through the coil.

If the switch is closed, a current can flow in the coil. By Lenz's and Faraday's Laws, the magnetic flux in the coil is changing, and so a current will flow to oppose the change. This current will provide a force that opposes the motion of the magnet, so the magnet will move more slowly than it would if the switch were open.

c As it stands, the device will give a flash of light while the magnet is moving in the coil, then stop, so shaking it would give a pulsing light. If we incorporated a circuit including a battery, we might be able to charge the torch up by shaking it and then use it for a while before charging it up again. This would make it easier to see things because the torch would not be flicking on and off and being shaken all the time!

3 a As we move the metal in the magnetic field, the electrons in it experience a force, just like those in a coil in a generator. Unlike electrons in a coil, they can flow wherever they like, not just along wires, because the disc is relatively large. Because the magnetic force is always perpendicular to the velocity of the charge and to the field, the electrons tend to move in circular paths, giving rise to eddy currents (shown in the question diagram).

The kinetic energy of the disc (from pedalling) is now being used to create these eddy currents. The circulating eddy currents will create magnetic fields that will oppose the change in flux that created them. Therefore, the wheel slows down and the rider feels resistance. If more of the disc is subject to magnetic field, more eddy currents will be created and the resistance increases.

b A saucepan with a thick metal base is placed above a changing magnetic field created by an electromagnet with an AC current source. Eddy currents will be induced in the base of the saucepan. The metal of the saucepan has some electrical resistance so it heats up as the current flows in it. This heat can be used to cook.

4 If we have a base or pad with a coil in it, and the coil is connected to an AC voltage – such as that in the mains – then the coil will produce a time-varying magnetic field. A second coil inside the device, sitting on top of the first, will experience the changing magnetic field, and the changing flux will cause a current to flow. That current can be used to charge a battery connected to the second coil.

5 A changing magnetic field causes a current in a nearby conductor. That current generates its own magnetic field that opposes the change in the driving field. If the second field added to the first, rather than opposing it, the two fields would increase together, each making the other larger, without extra input of energy. This violates conservation of energy.

Because the back emf is a consequence of Lenz's Law, with the field that drives the motor operating on the coil in the motor, the back emf can be viewed as a consequence of conservation of energy.

MODULE SIX: CHECKING UNDERSTANDING PAGE 63

1 D

$$F = qE = ma$$

$$m = \frac{qE}{a}$$

$$= \frac{(5.0 \times 10^{-9})(26\,000)}{0.38}$$

$$= 3.4 \times 10^{-4}\,\text{kg} = 0.34\,\text{g}$$

2 **B**

$W = Fd\cos\theta$, but the angle between the force (given by $F = qvB\sin\theta$) and the direction of motion is 90°, so work comes out to be zero.

3 **D** All of the above.

4 **A** The induced emf between any two points in a changing magnetic field is in fact *not* independent of the path between the two points. It will depend on the area enclosed by the path, for example.

5 **A** a split ring commutator

6 $I_1 = 3I_2$, so

$$\frac{F}{\ell} = \frac{\mu_0 3I_1^2}{2\pi r}$$

$$I_1^2 = \frac{2\pi r}{3\mu_0}\frac{F}{\ell}$$

$$I_1 = \sqrt{\frac{2\pi(0.012)(6.3 \times 10^{-5})}{3(4\pi \times 10^{-7})}}$$

$$= 1.12\text{ A}$$

7 a ε: the induced emf

Φ: the magnetic flux (through the area or loop)

Δ: 'change in'; for example, Δt is the change in time.

b   Lenz's Law: The induced emf (and any resulting current) is such that it opposes the change in flux that induced it.

MODULE SEVEN: THE NATURE OF LIGHT

REVIEWING PRIOR KNOWLEDGE PAGE 65

1 In longitudinal waves, the oscillations are parallel to the direction of propagation of the wave. In transverse waves, the oscillations are perpendicular to the direction of propagation of the wave.

2 A 2, B 4, C 1, D 3

3 a $v = f\lambda$. As a radio wave is part of the EM spetrcum, v becomes c and we get

$$\lambda = \frac{c}{f} = \frac{3 \times 10^8}{6 \times 10^8} = 5.0 \times 10^{-1}\text{ m}.$$

b $f = \frac{1}{T} \therefore T = \frac{1}{f}$ so $\frac{1}{6 \times 10^8} = 1.67 \times 10^{-9} = 1.7 \times 10^{-9}\text{ s}$

4 The refractive index of a material, given by the relationship $n_x = \frac{c}{v_x}$, and Snell's Law, which gives the relationship between the incident and reflected rays $n_1\sin\theta_1 = n_2\sin\theta_2$.

5 $f' = f\frac{(v_{wave} + v_{observer})}{(v_{wave} - v_{source})} \therefore 1230\left(\frac{340 + 0}{340 - 34}\right) = 1323.48$ so frequency observed is 1323 Hz.

Frequency change $= f' - f = 1323 - 1230 = 93\text{ Hz}$

Chapter 8: Electromagnetic spectrum

WS 8.1 PAGE 66

1 Pythagoras – light travels from the eye to the object; Euclid – light travels in straight lines; Hasan Ibn al-Haytham – light is reflected off objects into the eye, allowing us to see; René

Descartes – there are tiny spheres, called plenum, that allow for light to travel through them like sound waves; Isaac Newton – light is made of lots of little particles, called corpuscles; Christiaan Huygens – light travels in waves

2 a   That electric field lines start at a positive charge and end at a negative charge; also that the electric flux leaving an enclosed space is proportional to the charge within it.

b   That magnetic fields are continuous, and therefore monopoles cannot exist. This also states that the net magnetic flux through a surface must equal zero.

c   That a changing magnetic field will induce an electric field and an emf in a conductor/coil/wire that will oppose the change that caused it (Lenz's Law).

d   That magnetic fields can be produced by a moving charge as well as by a changing the electric field.

3 a   Using his equations, Maxwell formed a wave equation that gave him the following expression for the speed of light:

$$v = \frac{1}{\sqrt{\mu_0\varepsilon_0}}.$$

Knowing the values for μ_0 and ε_0, from earlier research, he was able to show that $v = 3.1 \times 10^8\text{ m s}^{-1}$, which was close to the accepted speed of light at that time.

b   Because the waves are produced by an oscillating charge, they will have the same frequency as the oscillations of the charge, therefore allowing the prediction that, if you change the rate of the oscillations, the frequency of the emitted wave will also change.

4 By making a prediction, based on previous research, scientists can design experiments that will test these predictions as either correct or not, which will provide support (or not) for the associated theory. By having the prediction in the first place, it will drive the quest for scientific knowledge forward.

WS 8.2 PAGE 68

1 a E: the electric field

B: the magnetic field

b   When a charged particle oscillates it will induce a magnetic field, as per Ampere's Law. This in turn will induce an electric field, as stated by Faraday's Law. When the two fields are produced, they will cause energy to be carried away from the source by the waves.

2 Maxwell predicted that electromagnetic waves:

- were made of oscillating electric and magnetic fields oriented perpendicularly
- were self-propagating
- were created by oscillating charged particles
- moved at a speed now known as the speed of light (300 000 000 m/s)
- could have a range of frequencies and wavelengths whereupon the visible spectrum comprised just one small section.

3 Answers will vary but could include the following information within each area.

Radio waves: rapid oscillations of electrons; antennas and other receiving equipment; uses include communication

Microwaves: circular motion of electrons controlled by magnetic or electric fields such as a magnetron; diodes or radio antennas tuned to high RF frequencies; uses include food preparation

Infrared light: vibrating molecules and other atoms; specialist infrared photo-sensing equipment, heat-sensing receptors, thermopiles; uses include heat lamps for tissue therapy

Visible light: excited electrons falling to their rest state within atoms; human eyes, other light-recording devices such as cameras; uses include vision

Ultraviolet light: excited electrons in an atom or molecule falling to lower energy levels; photoelectric cells, specialised photographic film; uses include solar cells

X-rays: electrons that are accelerated to very high speeds, then collided with a metal target; specialised photographic films, Geiger counters; uses include medical diagnostics

Gamma rays: when a nucleus undergoes radioactive decay; scintillation tubes, specialised medical equipment; uses include radiotherapy and food sterilisation

WS 8.3 PAGE 70

1

Scientist	Year	Experiment	Measurement in $km\,s^{-1}$
Galileo	1638	Lanterns on hills	Instantaneous
Romer	1676	Eclipse of the moon Io	220 000
Bradley	1728	Stellar aberration	301 000
Fizeau	1849	Toothed wheel	315 000
Weber and Kohlrausch	1856	Measurements of μ_0 and ε_0	310 000
Foucault	1862	Rotating mirror	298 000
Michelson	1926	Rotating mirror	299 796 ± 4
Froome	1958	Radio interferometry	299 792.5 ± 0.1
Woods	1978	Wavelengths and frequencies of lasers	299 792.459 ± 0.001

2 Answers will vary, but they should include some of the following information:

Early thoughts were that the speed of light was infinite.

Early measurements used astronomical measurements because the technology of the day allowed for this. These experiments demonstrated that the speed of light was finite.

The first experimental evidence was shown by Fizeau, with the toothed wheel experiment.

Fizeau's ideas were built on by Foucault due to improvements in technology that allowed for greater accuracy of the spinning mirror. This was again improved on by Michelson, who increased the distance to get more accurate results.

The development of technology allowed for Rosa and Dorsey to more accurately measure μ_0 and ε_0, which helped to strengthen Maxwell's prediction of the speed of light.

The current measurement is given by the hyperfine frequency of caesium, which defines the second and the metre as how far light will travel in this time. This comes to $299\,792\,458\,m\,s^{-1}$.

All the experiments should have pros and cons to them.

3 a Fong knows that the microwaves produced are part of the electromagnetic spectrum and therefore should travel at the speed of light. If he gets the frequency of the microwaves from the manufacturer or microwave labelling and then measures the distance between the hot spots, he will have half of the wavelength due to a microwave oven being a resonant cavity. Then, using $c = f\lambda$, he should be able to work out the speed of the waves.

b It doesn't say if Fong removed the turntable from the microwave. Having the turntable in place would make the measurements invalid because the marshmallows would be turning and this would cause the hot spots to be too large to measure. When he is measuring the hot spots he would not be able to find the definite centre, which could weaken the accuracy of the wavelength measurement.

WS 8.4 PAGE 72

1 Continuous spectra – formed by the splitting of white light, often using a prism or some other glass-type object that will refract the light. The full continuous spectrum is visible, showing all the colours of a rainbow, so it will look like a rainbow.

Emission spectra – caused when an electron in a gaseous atom that has been excited, due to absorbing energy by an electric discharge or flame, releases electromagnetic radiation when falling back down to a lower energy level. Because the emitted electromagnetic radiation corresponds to a specific energy/frequency/wavelength, the resulting spectrum constitutes discrete emissions at these frequencies that are specific to the element(s) involved.

Absorption spectra – caused when a continuous spectrum passes through a cool cloud of gaseous elements. Certain photons interact with and excite electrons in the atomic elements, causing them to move to a higher energy level. The resulting spectrum would be similar to a continuous spectrum but with black lines corresponding to the wavelength of the photons that have interacted with the electrons. These are specific to the atoms involved.

2 a Sample 1: Lithium

b Sample 2: Hydrogen

c Sample 3: Beryllium

d Sample 4: Helium

e Sample 5: Hydrogen and lithium

f Sample 6: Lithium, hydrogen and beryllium

3 a As a star is a black body and emits all frequencies of the electromagnetic spectrum, you can relate the peak wavelength emitted to the temperature through the relationship $\lambda_{max} = \dfrac{b}{T}$. This means that you can work out the surface temperature of the star from its apparent colour.

b The top spectrum has moved towards the red end of the visible light spectrum. This would indicate that the wavelengths are getting longer, and the star is 'red shifting' or moving away from the observer. The bottom spectrum has moved towards the blue end of the visible light spectrum and its wavelengths are getting shorter, so it is 'blue shifting': this would indicate that the star is moving towards the observer. If the star is rotating, you would see both red and blue shift at the same time. By analysing the broadening of the spectral lines the rotational velocity can be calculated.

c As the photons are released from the core of star, they interact with the gas that surrounds the core. The denser this gas is, the more interactions the photons will have. This will lead to broader lines in the spectrum. With larger stars the effect of gravity on the outer layers is smaller than the effect on layers closer to the core. This leads to less dense gas in the outer layer and fewer interactions, which in turn leads to finer lines in the emission spectrum. So, a larger star will have finer spectral lines.

Chapter 9: Light: wave model

WS 9.1 PAGE 75

1 Newton proposed that light came from a source in the form of corpuscles, which were perfectly elastic, weightless and

~continued in right column ▲

rigid particles. Like other particles, corpuscles travelled in straight lines and would obey the laws of inertia. Different colours were due to different masses of the corpuscles.

2 Huygens stated that light travels in waves that leave a source at a uniform velocity and in all directions. Any point on the wavefront could act as the point source for a new wavelet, propagating in the direction of travel. To allow for the light to travel through a vacuum, Huygens proposed that there was a medium in the vacuum called the lumeniferous aether. He suggested that different colours of light were due to the different wavelengths emitted from the source.

3

	Newton	Huygens
Reflection	As the light is made up of many tiny particles, they will bounce off a surface with the same speed and angle at which they approached.	When the wave front meets a surface, new wavelets are formed at the interface, producing a wavefront that will then propagate.
Refraction	The light particles experience a force from the particles in the medium, causing them to bend towards the normal and travel at a faster speed.	When one point of the wavefront hits the refractive surface it forms a new wavelet that will move at a different speed to the original wave, causing it to change direction. This will obey Snell's Law and shows that light will slow down in the denser medium.
Diffraction	Cannot be explained by this model.	Diffraction is the bending of waves due to an opening. The smaller the opening, the larger the diffraction of the wave.
Rectilinear propagation	As the corpuscles are tiny particles moving at high speed, they travel in straight lines away from the source.	Cannot be explained by this model.

4 Answers should include the information below.

Evidence for Newton's model: the success of geometric optics, indicating that light travels in straight lines, and how this could explain the phenomena of reflection and refraction.

Newton incorrectly stated that light would speed up in a denser medium. This was accepted until 1850, when Foucault showed that light slowed down in water when compared to air. Newton had to modify his model to explain double refraction of light, and he could not explain diffraction of light with his model.

Huygens' principle could explain the behaviour of light as a wave and that light showed many of the behaviours associated with waves. The fact that he could not explain

~continued in right column ▲

rectilinear propagation was overlooked, since diffraction and interference could be explained using his wave model. Some of the issues with Huygens' model arose because it was based primarily on light pulses as opposed to periodic waves. This was due mainly to his thinking that light was a compressional wave such as a mechanical wave.

While both models explained most of the behaviours of light, neither could explain all the behaviours and so were both incomplete. They also had some aspects which were later shown to be incorrect, such as Newton's prediction that light would speed up in a denser medium. However, Newton's model was more widely accepted due to his higher standing in the scientific community compared with Huygens.

5 a

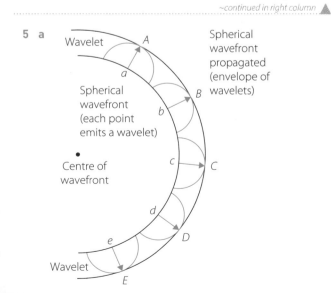

b

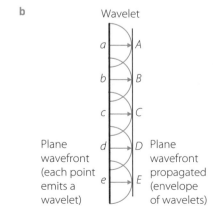

1 Thomas Young set up his experiment to investigate whether light was a wave, because a wave would produce a diffraction pattern, which is a unique property of a wave. His experiment set up a point source of light going through a small pinhole. This allowed for the light to diffract when moving through the pinhole and reach two slits on another sheet a short distance away, at the same time, as per Huygens' principle. He set it up in this way as it was the best way to have a coherent light source on the two slits. Today we would use a laser to produce the coherent light source. As the light passed through these slits it diffracted and created two new wave fronts. If the waves interfered constructively, then a light spot would appear on the screen at the antinode of the wave. If the waves interfered destructively, then a dark spot would appear on the screen, at the node of the wave.

As Young observed a diffraction pattern with the experiment, it can be said that light certainly acts as a wave.

2

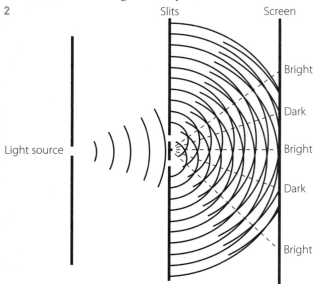

3 Monochromatic light – as you require the same wavelength (colour) of light to cause the interference pattern. Light of more than one colour would contain different wavelengths so it would not interfere as clearly.

Coherent light source – as the light and dark spots are reliant on having the waves either in phase or half out of phase, the light needs to be leaving the slits at the same time.

4

The light from the laser would show a diffraction pattern on the screen relative to the double slit, due to the light diffracting through the double slit. This pattern would have a central bright spot with several bright and dark lines spreading out and becoming less intense. This is due to the intensity of the light being greatest when the angle is 0°.

5 a As the angle can be related to the height above where the normal hits the screen (y) and the distance from the slit to the screen (L), this can be given as $\sin\theta = \dfrac{y}{L}$.

Substituting into the double slit formula we get:

$$\frac{dy}{L} = m\lambda \therefore y = \frac{m\lambda L}{d} = \frac{1 \times (696 \times 10^{-9}) \times 2}{5 \times 10^{-4}}$$
$$= 2.784 \times 10^{-3} = 3 \times 10^{-3}\,\text{m}$$

b $d\sin\theta = m\lambda$

$$\therefore \theta = \sin^{-1}\left(\frac{m\lambda}{d}\right) = \sin^{-1}\left(\frac{1 \times 696 \times 10^{-9}}{5 \times 10^{-4}}\right)$$
$$= 0.079\,75 = 0.08°$$

6 Using Huygens' principle, as the light waves diffract at the edge of the disc, they will start two new wave fronts that will interfere constructively and create a bright spot where they hit the screen.

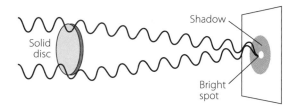

7 a These are suggested answers. Other answers may be acceptable.

Risk	Risk level	Mitigation
The light bulb becoming very hot if it is left on for a long time	Medium	Have the light on only when making the required measurements.
Being hit with the ruler	Low	Put the ruler away between uses.
Breakage of the light bulb	Low	Ensure that the light is held securely and that if it does break the glass is cleared up immediately.

b By using an incandescent bulb, the students would have the light going in all directions. This would affect their ability to be able to make accurate measurements of the distances between the maxima as the room would not be dark enough. The students would need to work in a darkened room and use a laser pointer to provide a concentrated beam of light or create a hood for the bulb so that they could direct the light in one direction. The light from the incandescent bulb may still not be strong enough to get clear distinction between maxima and so using a laser would be the more effective way of carrying out this experiment. Another problem with the incandescent bulb is that they would not be using a monochromatic light source, so they would have a dispersion pattern as well as a diffraction pattern.

8 a The single-slit and double-slit experiments would both have a central maximum, which is the brightest spot, with less intense maxima evenly spaced away from the central maximum. The single slit bright spot would be wider, with the intensity dropping off very quickly from this point, while the double slit would have a more even spread in intensity. This is due to the slit width of a single-slit pattern and the interference caused by Huygens' principle. Huygens' principle dictates that each point on the wave front acts as a new point source for a wave.

The interference pattern caused by the double slit is the result of the slit width and slit separation. Diffraction gratings are similar to double slits in that they both show Fraunhofer diffraction and multiple waves interfering with each other. The pattern produced by the diffraction grating is much sharper due to the maxima immediately dropping in intensity to the minima so that you see the bright spots clearly. The width of the maxima is inversely proportional to the number of lines in the grating.

b   i   Double slit: $d\sin\theta = m\lambda$ where d is the slit separation

   ii   Diffraction grating: $d\sin\theta = m\lambda$ where d is the grating separation

   iii   Single slit: $a\sin\theta = m\lambda$ where a is the slit width

WS 9.3 PAGE 81

1   The light intensity will drop by 50%. This is because the filter can be aligned in one of two possible ways, horizontal or vertical, therefore blocking half the light.

2   a   **1**   Set up a light source 1 m away from the light meter.

     2   Set up the first polariser between the light source and the meter and record the intensity.

     3   Set up the second polarising filter between the first filter and the meter. Ensure that the polarisers are set up in a way that allows all of the polarised light through.

     4   Move the second filter through 10° increments, recording the intensity as it is moved through to 180°.

b

Angle between filter 1 and filter 2 (°)	Light intensity through the first filter (W m^{-2})	Light intensity through the second filter (W m^{-2})
0	250	250
10	250	242.5
20	250	220.8
30	250	187.5
40	250	146.7
50	250	103.29
60	250	62.5
70	250	29.24
80	250	7.54
90	250	0
100	250	7.54
110	250	29.24
120	250	62.5
130	250	103.29
140	250	146.7
150	250	187.5
160	250	220.8
170	250	242.5
180	250	250

c

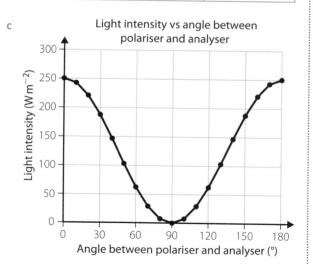

Light intensity vs angle between polariser and analyser

d   The results follow a $\cos^2$ graph, which is what you would expect from Malus' Law. Visually, the light intensity coming through the filters drops to zero intensity as the angle approaches 90° and then increases after that, again supporting Malus' Law.

e   The investigation was valid because all the variables were controlled except for the angle between the two filters, which was the independent variable. This shows that the only reason for the change in intensity between the filters is due to this one variable. The experiment would need to be repeated several times with the data points averaged out for it to be considered reliable.

3   a   90° to each other

b I_{max} from the first filter would be halved, so $1500\,\text{W}\,\text{m}^{-2}$ would be transmitted through the first filter.

Using Malus' Law for the second filter, we would have $I = I_{max}\cos^2\theta$.

$I_2 = 1500\cos^2 45° = 750\,\text{W}\,\text{m}^{-2}$

$I_3 = 750\cos^2 45° = 375\,\text{W}\,\text{m}^{-2}$

Therefore adding the third filter increases the transmitted light from 0 to $375\,\text{W}\,\text{m}^{-2}$.

4   $I = I_0\cos^2\theta$ so $42\%\,I_0 = I_0\cos^2\theta$

$0.42 I_0 = I_0\cos^2\theta$

$0.42 = \cos^2\theta$

$\cos\theta = \sqrt{0.42}$

$\theta = \cos^{-1}\left(\sqrt{0.42}\right) = \cos^{-1}(0.648\,074)$

$= 49.60°$

5   $I_1 = I_0\cos^2\theta$ so $I = I_0\cos^2 45° = 0.5 I_0$

$I_2 = I_0\cos^2\theta$ so $I = I_0\cos^2 30° = 0.75 I_0$

Percentage increase $= \dfrac{I_2 - I_1}{I_1} \times 100 = \dfrac{0.75 - 0.50}{0.50} \times 100$

$= 50\%$

Chapter 10: Light: quantum model

WS 10.1 PAGE 84

1   A black body radiator is a theoretical body that absorbs all incident EM radiation regardless of frequency or angle. This means that it will also emit all wavelengths as a function of temperature and irrespective of size or composition.

2   The ultraviolet (UV) catastrophe came about due the attempted application to electromagnetic waves of classical physics laws describing wave behaviour. Classical physics predicted that, at higher temperatures, emissions at higher frequencies would be increasingly prevalent and tend towards infinity. This led to the creation of the Rayleigh–Jeans Law for how light would behave in relation to temperature. The law was accurate for the low energy long wavelengths of light; however, this law implied that there would be infinite energy at shorter wavelengths. As this would violate the law of conservation of energy, as well as not being seen in the experimental data, this led to the coining of the term 'UV catastrophe' to describe the failure of the classical model. The graph shows the discrepancy between observation and the predictions of the classical model.

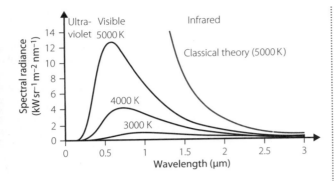

3 a $\lambda_{\max} = \dfrac{b}{T} = \dfrac{2.898 \times 10^{-3}\,\text{m K}}{30\,000\,\text{K}} = 9.66 \times 10^{-8}\,\text{m}$

b $\lambda_{\max} = \dfrac{b}{T} \therefore T = \dfrac{b}{\lambda_{\max}} = \dfrac{2.898 \times 10^{-3}\,\text{m K}}{828 \times 10^{-9}\,\text{m}} = 3500\,\text{K}$

4 That the particles that were oscillating within the black body radiator could only have specific, discrete values of energy (that is, one value or the next, but not the values in between). This means that energy is quantised.

That if the oscillators could only absorb specific quanta or chunks of energy then they could also only emit the same specific quanta, which had to be integer multiples of hf.

5 If the discovery goes against the current thinking about what is happening in the natural world, then it can cause the scientist to think that the findings are incorrect, or that they have interpreted them wrongly. Planck was able to describe experimental results mathematically by introducing the notion of quantisation. He did this because of the experimental results, but without a theoretical basis. A lack of theoretical foundation could lead to bias being brought into the investigation, which would influence their results. However, in this case Planck was correct, as was proved many years after the event, and led to the development of quantum mechanics.

6 $E = hf = (6.626 \times 10^{-34}) \times (8 \times 10^{14})$
  $= 5.3 \times 10^{-19} = 5 \times 10^{-19}\,\text{J}$

7 $E = hf$ so $f = \dfrac{E}{h}$

  $f = \dfrac{4.3 \times 10^{-18}}{6.626 \times 10^{-34}} = 6.49 \times 10^{15}\,\text{Hz}$

8 $E = hf = (6.626 \times 10^{-34}) \times (7.6 \times 10^{16}) = 5.04 \times 10^{-17}\,\text{J}$

  $= \dfrac{5.04 \times 10^{-17}}{1.602 \times 10^{-19}} = 315\,\text{eV}$

WS 10.2 PAGE 86

1 Answers should follow the following timeline.

Henrich Hertz first showed evidence of the photoelectric effect in 1887, when he showed that electricity flowed more easily when certain light was shone on a polished metal surface.

Philipp Lenard in 1902 showed that the maximum velocity of a particle ejected from a metal surface was independent of the intensity of the light.

Albert Einstein in 1905 proposed that this energy was related to Planck's equation. This matched Planck's 'mathematical trick' of $E = hf$.

This was proven by Millikan in 1916, who showed Einstein's work matched Planck's.

In 1921 Einstein was awarded the Nobel Prize for this work.

2 a The minimum amount of energy required for an electron to 'escape' from a metal. This energy would be extracted from the total energy of the incident photon and would determine the maximum kinetic energy of the ejected electron.

b The minimum frequency of a photon required to cause an electron to be ejected from the surface of a metal.

3

Theoretical prediction using the wave model of light	Experimental evidence
Electrons should be emitted at all frequencies of light.	Electrons are only emitted when the light is above the threshold frequency.
At low intensities there should be a delay between the light hitting the metal surface and electrons being emitted.	Electrons are emitted immediately when the light is incident on the metal surface.
The current produced should be dependent on both the intensity and the frequency.	The current produced is dependent only on the intensity of the light due to the ratio of one photon to one electron.
Maximum kinetic energy should be related to the intensity of the light and not the frequency.	The maximum kinetic energy is only related to the frequency of the incident light and not the intensity.

4

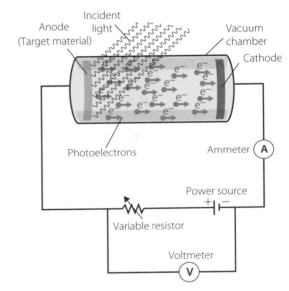

Anode – metal surface that will emit electrons when incident light above the threshold frequency hits it

Cathode – positively charged metal plate that attracts the electrons that have been emitted

Vacuum chamber – ensures that the electrons move across the potential divide without interacting with any other matter

Photoelectrons – surface electrons that have been emitted from the metal surface of the cathode

Ammeter – to register the current produced by the photoelectric effect

Voltmeter – to register the voltage in the circuit

Power source – to create a potential difference between the cathode and anode so that the photoelectrons will flow across the gap, creating a current

Variable resistor – to change the resistance to control the voltage in the circuit so that the stopping voltage can be measured

9780170449687

5 a To find the frequency, use $c = f\lambda$: the energy of the ejected photon is equal to the stopping voltage applied. Note that the energy can also be given in joules.

Wavelength (nm)	Stopping voltage (V)	Energy (eV)	Frequency (Hz)
166	−4.49	4.49	18×10^{14}
187	−3.87	3.87	16×10^{14}
214	−3.00	3.00	14×10^{14}
250	−2.12	2.12	12×10^{14}
300	−1.37	1.37	10×10^{14}

b

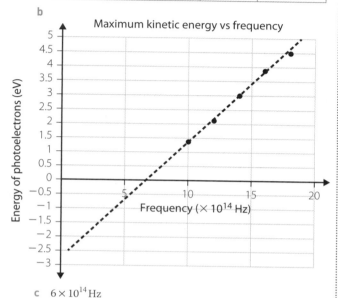

Maximum kinetic energy vs frequency

c 6×10^{14} Hz

6 Work function is $\phi = hf_0 = 6.626 \times 10^{-34} \times 6 \times 10^{14}$
$= 3.98 \times 10^{-19}$ J $= 2.48$ eV

Note the vertical intercept can be used for this as well.

7 $K_{max} = V_{stop}q = V_{stop}e$
4.49 V $\times 1.602 \times 10^{-19}$ J $= 7.19 \times 10^{-19}$ J

8 $K_{max} = \frac{1}{2}mv^2 = V_se$

$\therefore v = \sqrt{\frac{2V_se}{m}} \therefore \sqrt{\frac{2 \times 4.49 \text{ V} \times 1.602 \times 10^{-19} \text{ J}}{9.109 \times 10^{-31} \text{ kg}}}$

$= 1.2567 \times 10^6 = 1.26 \times 10^6$ m s^{-1}

9 Speed from shortest wavelength 1.26×10^6 m s^{-1}, from Question 8.
Energy from longest wavelength:

$K_{max} = V_{stop}q = V_{stop}e = 1.37 \times 1.602 \times 10^{-19} = 2.19 \times 10^{-19}$ J

$K_{max} = \frac{1}{2}mv^2 = V_se$

$\therefore v = \sqrt{\frac{2V_se}{m}} = \sqrt{\frac{2 \times 1.37 \text{ V} \times 1.602 \times 10^{-19} \text{ J}}{9.1 \times 10^{-31} \text{ kg}}}$

$= 6.95 \times 10^5$ m s^{-1}

Therefore, the shortest wavelength causes the emitted electrons to be travelling nearly twice as fast as the electrons emitted by the longest wavelength.

10 Gradient is $\dfrac{\text{rise}}{\text{run}} = \dfrac{4.49 - 1.37}{18 \times 10^{14} - 10 \times 10^{14}} = 3.9 \times 10^{-15}$ eV s

$= 6.25 \times 10^{-34}$ J s

The gradient should show Planck's constant and, as this result is very close, the investigation was a success. Note other points along the gradient can be used and should give the same result.

11 The gradient should be the same for each metal because this was shown to be Planck's constant. However, the threshold frequency and work function value depend on the metal, so will be higher or lower depending on the different amounts of energy required to release the electrons.

12 The results show that the maximum kinetic energy of a released electron is proportional to the difference of the frequency of the incident light and the threshold frequency for the metal. The intersection of the line with the x-axis is the threshold frequency of the metal: below this frequency no electrons will be emitted from the metal.

13 a The greater the intensity of light on a metal, the greater the current produced will be.

b No. The students are measuring two variables because they change the wavelength as well as the aperture size. The investigators would need to measure only the aperture size without changing the colour filter.

c The backing voltage of the experiment would need to be set at zero for every trial.

14 a

Diameter of aperture (mm)	Area of aperture (mm²)	Current (mA)
7	38.4	0.10
10	78.6	0.17
14	153.9	0.22
20	314.2	0.26

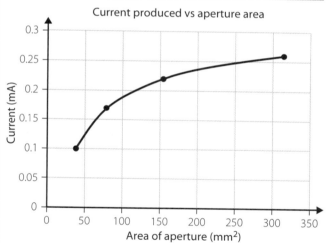

Current produced vs aperture area

b The graph shows current increasing as the size of the aperture increases. This would indicate that if the intensity of the light is increased then the current will increase as well. The graph starts to level off, which would suggest that, according to these results, the current will reach a point from which it will not increase regardless of the intensity of light.

15 The investigation showed that if the intensity of light increases the emitted current also increases. This supports the hypothesis.

16 Answers may vary but must include the following:

As light is quantised it will give only a certain amount of energy to each electron that it interacts with.

There is a certain amount of energy required for an electron to be ejected from the surface of a metal. This is known as the work function.

The energy given to the electron is defined by the frequency or the wavelength of the incident light.

The maximum kinetic energy of the ejected particle is given by the photoelectric effect equation $K_{max} = hf - \phi$, which accounts for the total energy and therefore shows energy is conserved.

Chapter 11: Light and special relativity

1 a

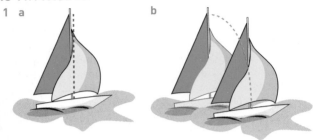

b

2 In diagram **a**, the ball has moved straight down due to the observer being in the same frame of reference as the ball and ship. In diagram **b**, because the observer is viewing it from outside the frame of reference, they see the ball moving at a certain speed forwards in addition to the vertical motion and this will give the impression that the ball has followed a parabolic path to the deck.

3 As long as the boat is moving at a constant velocity or is stationary, then both frames of reference would agree that the laws of motion are the same in both frames, and therefore both frames of reference are correct.

4 a If you are not accelerating, there are no experiments that you can carry out to find out if you are inside a moving or a stationary frame of reference.

b That the laws of motion are the same for all inertial frames of reference, that the conservation laws of energy and momentum apply in inertial frames of reference, and that all inertial frames of reference are equally valid.

5 The aether would constitute an absolute frame of reference for Earth because it would be a non-moving medium with everything else moving through it. This goes against the idea that there is no absolute frame of reference. It also would imply that the speed of light would be different depending on the direction of travel within the aether.

6 a

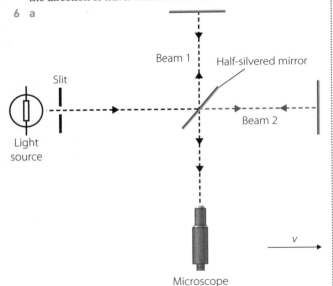

The Michelson and Morley experiment of 1887 showed that the speed of light was constant in opposite directions of the interferometer. This was consistent with the idea that the speed of light was a constant.

b This helped to support Einstein's postulate that the speed of light is consistent in all frames of reference.

7 The experiment was both a valid and reliable experiment that supported the null hypothesis. It was reliable for the following: repeated many times over different years, rotated the apparatus through 90° to test both possible directions of the aether, tested by waiting for Earth to be in different parts of its orbit and then tested again.

It was valid due to all other variables being controlled (such as mercury bed for apparatus, performed at same height above sea level), based on sound science of the day, and it tested the aim and hypothesis.

8 That the laws of physics are the same in all inertial frames of reference.

That the speed of light is constant in all inertial frames of reference independent of the motion of the source or the observer.

9 a Relative motion predicts that the gamma rays would be travelling at the speed of light plus the speed of the subatomic particles, therefore travelling faster than the speed of light in front of the particles and at significantly less than the speed of light behind the particles.

b The speed of light is constant for all inertial observers irrespective of motion of source, so the gamma rays (part of the EM spectrum) would travel at c both in front of and behind the particle no matter the motion of the observer.

10 There is no current evidence against Einstein's postulates. However, as with all scientific ideas, this does not mean that we stop looking for some. If any evidence is found to oppose the ideas of the postulates, then the Theory of Special Relativity will need to be reconsidered.

WS 11.2 **PAGE 95**

1 More accurate and reliable forms of measurement were required. Physical objects undergo changes, such as wearing down, expansion and contraction. To use in different locations, copies and prototypes need to be maintained. The use of the speed of light, a natural phenomenon with a precise value, provides more accurate and reliable measurements of distance.

2 A thought experiment is used to extend and explore the outcomes of a theory. It is carried out when it is not feasible to conduct the physical experiment. The results produced are valid, because the logic and reasoning of the thought experiment are consistent with and drawn from scientific theories.

3 As discussed above, a thought experiment is a non-empirical way to explore the consequences of theories and their assumptions, while a paradox is a seemingly contradictory situation that, despite sound theory, can lead to a logically unacceptable conclusion.

9780170449687

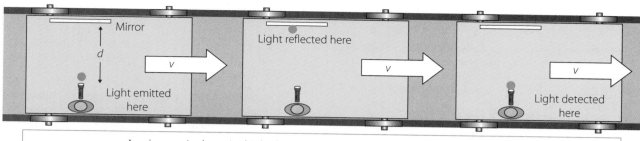

4

Mirror

d

v

Light emitted here

Light reflected here

v

v

Light detected here

An observer in the train thinks that the light has travelled a distance 2d in a time t

v

Light emitted here

d

v

Light emitted here

v

2s

Observer on the bank

This means that the observer outside the train would see the light take a much longer path and would therefore conclude that the light had taken longer to make the same journey since it must always travel at the same speed.

5   The twin paradox: one twin goes away from Earth at close to the speed of light for a period of time and then comes back. The stay-at-home twin will be older than the travelling twin, but a naive application of time dilation would suggest their situations are symmetrical and each should expect the other to be younger. This is resolved because their situations are not symmetrical: the travelling twin has been in two separate inertial frames and has been in a non-inertial frame (slowing, speeding up and turning), so we cannot apply special relativity in this case.

6   Identify variables: t_0 = astronaut onboard ship, therefore 30 s; t = observer outside the spaceship; $v = 0.75c$. To make the

equation a bit simpler we can take the ratio $\frac{v^2}{c^2}$ as 0.75c.

$t = \dfrac{30}{\sqrt{1 - 0.75c^2}} = 45.4\,\text{s}$

7   Identify variables: $t = 3\,\text{h} = 10\,800\,\text{s}$, $\frac{v^2}{c^2} = 0.82c$, $t_0 = ?$

$t_0 = 10\,800\sqrt{1 - 0.82c^2} = 6182\,\text{s} = 1\,\text{h}\,43\,\text{min}$

8   $t = 10$ years, $t_0 = 6$ years, $v = ?$

$t = \dfrac{t_0}{\sqrt{1 - \dfrac{v^2}{c^2}}}$ so $\sqrt{1 - \dfrac{v^2}{c^2}} = \dfrac{t_0}{t}$.

Rearranging, $1 - \dfrac{v^2}{c^2} = \left(\dfrac{t_0}{t}\right)^2$ so $\dfrac{v^2}{c^2} = 1 - \left(\dfrac{t_0}{t}\right)^2$

and $\dfrac{v}{c} = \sqrt{1 - \left(\dfrac{t_0}{t}\right)^2}$.

Substituting and solving, $v = c \times \sqrt{1 - \left(\dfrac{6}{10}\right)^2} = 0.8c$.

9

$v\ (\text{m s}^{-1})$	$\dfrac{v}{c}$	γ
3.00×10^2	1.00×10^{-6}	1
3.00×10^4	1.00×10^{-4}	1.000 000 005
3.00×10^7	0.100	1.005
1.40×10^8	0.467	1.131
2.00×10^8	0.667	1.342
2.50×10^8	0.833	1.809
2.90×10^8	0.967	3.906
2.99×10^8	0.997	12.26
2.999×10^8	0.9997	38.73

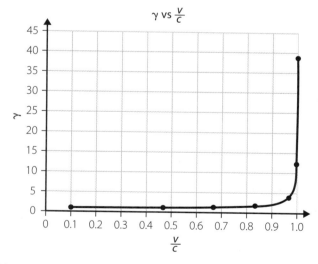

γ vs $\dfrac{v}{c}$

10 As you approach the speed of light the Lorentz factor will approach infinity. This will mean that the moving clocks will run increasingly slowly, tending towards time standing still.

11. The constant speed of light results in time dilation – two observers in different frames of reference record different times for the same event. They will, however, agree on their relative speed. The distance travelled during the event must therefore be recorded as different if the relative speeds are the same and the times for the event are different.

12. From the reference frame of the train, the train is longer than the tunnel because there is length contraction in the tunnel, which is moving relative to the train.

    However, to a stationary observer, the train is in motion and thus experiences length contraction in the direction of travel and it can fit into the tunnel.

13. Identify variables: $L_0 = 113.39 \times 10^9$ m, $L = ?$, $v = 0.65c$

    $113.39 \times 10^9 \sqrt{1 - 0.65^2} = 86.169 \times 10^9$ m

14. Identify variables: $L = 6.5$ m, $L_0 = 8$ m, $v = ?$

    $L = L_0 \sqrt{1 - \dfrac{v^2}{c^2}}$ so $\sqrt{1 - \dfrac{v^2}{c^2}} = \dfrac{L}{L_0}$.

    Rearranging, $1 - \dfrac{v^2}{c^2} = \left(\dfrac{L}{L_0}\right)^2$ so $\dfrac{v^2}{c^2} = 1 - \left(\dfrac{L}{L_0}\right)^2$

    and $\dfrac{v}{c} = \sqrt{1 - \left(\dfrac{L}{L_0}\right)^2}$.

    Substituting and solving, $v = c \times \sqrt{1 - \left(\dfrac{6.5}{8}\right)^2} = 0.58c$.

15. $L = L_0 \sqrt{1 - \dfrac{v^2}{c^2}}$ so $L_0 = \dfrac{L}{\sqrt{1 - \dfrac{v^2}{c^2}}}$.

    Substituting and solving, $\dfrac{532 \text{ light years}}{\sqrt{1 - 0.65^2}} = 700$ light years.

16. Muons are created during interactions in Earth's atmosphere approximately 10 km above the surface and they travel at 0.9996 of the speed of light. The half-life of a muon is 2.2 μs in a stationary frame of reference (such as in a laboratory), so using $d = vt$, the muons should decay before they reached the surface. (Distance travelled in a half-life = $(0.9996 \times 3 \times 10^8) \times (2.2 \times 10^{-6}) = 660$ m.) If we use the time dilation formula we get a relativistic time of

    $t = \dfrac{t_0}{\sqrt{1 - \dfrac{v^2}{c^2}}} = \dfrac{2.2 \times 10^{-6}}{\sqrt{1 - 0.9996^2}} = 77.79 \times 10^{-6}$ s. Thus while

    the muon experiences 2.2 μs elapsing in their stationary reference frame, we record a much greater period of time passing. If we put this new time into the first equation to find the distance covered by the muon, we get $(0.9996 \times 3 \times 10^8) \times (77.79 \times 10^{-6}) = 23\,327.7$ m. Thus, factoring in time dilation, it is clear muons can reach the surface of Earth.

17. a The clocks at the Observatory were the control clocks because they were in a stationary frame of reference relative to Earth. The clocks on the planes are in motion relative to the clocks on Earth and would be expected to record a different time for the event of the journey.

    b As the results fell within the predicted outcomes, they gave experimental evidence to support time dilation.

WS 11.3 **PAGE 99**

1. Rest mass is the only way that mass can be measured. It is the amount of matter that is in an object while it is at rest. Once an object is moving, the only way to measure the mass is through collisions.

2.

Velocity (m s^{-1})	Newtonian momentum (kg m s^{-1})	Relativistic momentum (kg m s^{-1})
3.0000×10^2	300	300
3.0000×10^5	300 000	300 000.15
3.0000×10^7	30 000 000	30 151 134.5
1.5000×10^8	150 000 000	173 205 080.8
2.0000×10^8	200 000 000	268 328 157.3
2.5000×10^8	250 000 000	452 267 016.9
2.9000×10^8	290 000 000	1 132 643 526
2.9995×10^8	299 950 000	1.16×10^{10}
2.9999×10^8	299 990 000	3.67×10^{10}

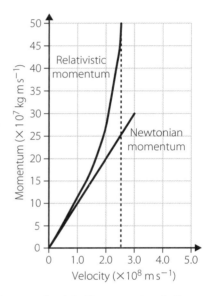

3. Kinetic energy is related to momentum in the equation

   $E_k = \dfrac{1}{2}mv^2$, which can be expressed as $E_k = mv \times \dfrac{v}{2}$ or

   $E_k = \dfrac{pv}{2}$. This shows that as the total momentum of an object increases then the kinetic energy must increase as well.

4. That mass could now be thought of as another form of energy, and therefore that the conservation of energy and conservation of mass were part of the same law.

5. a Difference between the mass of product and reactants:
   $\text{mass}_f - \text{mass}_i = 3.35 \times 10^{-27} - 2 \times 1.67 \times 10^{-27} = -1 \times 10^{-29}$ kg
   Mass cannot be negative so remove the negative sign and substitute in $E = mc^2$.
   $1 \times 10^{-29} \times (3 \times 10^8)^2 = 9.00 \times 10^{-13}$ J

b Change in mass from products to reactants
$$(2 \times 1.67 \times 10^{-27} + 6.64 \times 10^{-27}) - (2 \times 5.01 \times 10^{-27})$$
$$= -3.4 \times 10^{-29} \, kg$$
Substituting in $E = mc^2$:
$$3.4 \times 10^{-29} \times (3.0 \times 10^8)^2 = 3.06 \times 10^{-12} \, J$$

6 $m_e = 9.109 \times 10^{-31} \, kg$, so total mass is $2 \times 9.109 \times 10^{-31} \, kg$
$= 1.82 \times 10^{-30} \, kg$. Substititing into $E = mc^2$:
$$1.82 \times 10^{-30} \, kg \times (3.0 \times 10^8)^2 = 1.64 \times 10^{-13} \, J$$

7 For **5a**, the percentage of mass used was

$$\frac{\Delta m}{\text{mass of reactants}} = \frac{1 \times 10^{-29} \, kg}{2 \times 1.67 \times 10^{-27}} \times 100 = 0.18\%$$

For **5b**, $\dfrac{\Delta m}{\text{mass of reactants}} = \dfrac{3.4 \times 10^{-29} \, kg}{2 \times 5.01 \times 10^{-27}} \times 100 = 0.34\%$

For **6**, it is 100%, by definition.

**MODULE SEVEN: CHECKING UNDERSTANDING
PAGE 101**

1 A That light is an electromagnetic wave, created by oscillating electric and magnetic fields.

2 C The light will diffract with the red end diffracting the most.

3 B Where classical physics failed to explain experimental results at the ultraviolet part of blackbody radiation curves.

4 A In the direction of travel; the stationary observer

5 Answers can include GPS, particle accelerators, muons. Answers need to include that the object is moving with high speed, and the nature of the relativistic behaviour witnessed.

6 a $c = f\lambda$ so $f = \dfrac{c}{\lambda} = \dfrac{3 \times 10^8}{630 \times 10^{-9}} = 4.76 \times 10^{14} \, Hz$

b Energy of photon $= hf = 6.626 \times 10^{-34} \times 4.76 \times 10^{14}$

$$= \frac{3.15 \times 10^{-19} \, J}{1.6 \times 10^{-19} \, J \, eV^{-1}} = 1.97 \, eV$$

Energy is less than $\phi = 2.0998 \, eV$, so no electron is ejected.

c $v = \sqrt{\dfrac{2E_k}{m}} = \sqrt{\dfrac{2 \times 5.4 \times 1.602 \times 10^{-19} \, J}{9.109 \times 10^{-31} \, kg}} = 1.4 \times 10^6 \, m \, s^{-1}$

MODULE EIGHT: FROM THE UNIVERSE TO THE ATOM

REVIEWING PRIOR KNOWLEDGE PAGE 103

1 Protons have a positive charge and reside in the nucleus. Neutrons have no charge and also reside in the nucleus. Electrons have a negative charge and orbit the nucleus. Protons and neutrons have similar masses and are each about 1800 times heavier than an electron.

2 Natural radioactivity is caused by instability in the nucleus of an atom. This may be result from an imbalance in the ratio of protons to neutrons or from a large total number of protons.

3 Electric fields exert a force on all charged particles, in the direction of the field lines for positive charges and in opposition to the field lines for negative charges. Magnetic fields exert a force only on moving charged particles and this force is perpendicular to the magnetic field. The direction of the magnetic force is also dependent on the sign of the charge.

4 When there is relative motion between a source and an observer, the wavefronts being produced are either

compressed when the distance between the source and the observer is reduced or stretched when the distance is increased. Compressed wavefronts lead to a shorter wavelength and therefore higher frequency, whereas stretched wavefronts lead to a longer wavelength and lower frequency. This is easily observed in sound waves emitted from an emergency vehicle. As the vehicle approaches, the pitch of the sound wave is increased; once the vehicle passes, the pitch drops.

5 Light is considered to have both particle and wave properties.

6 Electromagnetic radiation has a relationship with energy that is dependent on frequency; that is, higher frequency photons have more energy than lower frequency photons. This can be expressed mathematically as $E = hf$.

7 $E = mc^2$ implies that mass and energy are two forms of the same thing – mass–energy – and that energy can be stored by creating mass or released by destroying mass.

Chapter 12: Origins of the elements

WS 12.1 PAGE 104

1

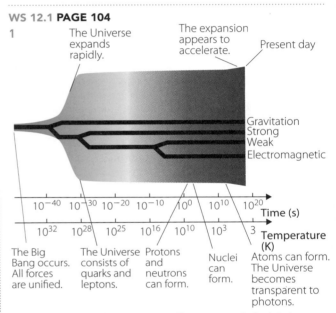

Note that the diagram is equally correct with the labels for the electromagnetic force and weak nuclear force interchanged.

2 Inflation

3 The radiation was continually absorbed and re-emitted by electrons (often at different frequencies) in different directions, scattering the light.

4 a The leftover gamma radiation (radiation that did not condense into matter) from the Big Bang is responsible for the CMBR.

b The expansion of the Universe after inflation has stretched out the wavelengths of the gamma radiation so much that they now correspond to the microwave region of the electromagnetic spectrum.

c The relative evenness of the temperature measured from the CMBR indicates that currently distant parts of the Universe were once close enough to exchange energy. This is direct evidence of an inflation period.

5 The decay of many unstable nuclei was observed to produce antimatter (see Chapter 15 for β^+ decay).

6 The weak nuclear force mediates the radioactive decay (such as beta decay) of nuclei, whereas the strong nuclear force is responsible for the binding of protons and neutrons within a nucleus.

WS 12.2 PAGE 107

1   Although the telescope was first used for astronomical purposes in the 17th century, our inability to see distant stars limited our understanding of the size of the Universe. Furthermore, our inability to measure distances in space beyond those calculated by parallax was a significant limitation.

2   By using the cepheid variable stars, Leavitt was able to determine distance beyond the Milky Way – a very significant step in determining both the size of the Universe and Hubble's confirmation of its dynamic nature.

3   Red shift and blue shift both refer to the Doppler effect as applied to visible electromagnetic radiation. Red shift occurs when the source and observer are moving away from each other, stretching the waves out and making the visible part of the spectrum appear more red. Blue shift occurs when the source and observer are moving towards each other, reducing the wavelength and hence making the visible part of the spectrum more blue in appearance. This relative motion may be caused by differing translational velocity or rotation in local settings but can also include the expansion of space-time between two objects within the Universe.

4   Hubble's observations showed that objects furthest from Earth are receding at the fastest rate and those closest to us recede the slowest. This suggests that tracing back in time would place all of these objects in close proximity at time = 0. The rate of expansion also allows for a determination for the age of the Universe. If the Universe was static, it would be impossible to calculate its age. Both the confirmation of a dynamic Universe and the estimation of the age of the Universe support the Big Bang Theory.

5   a

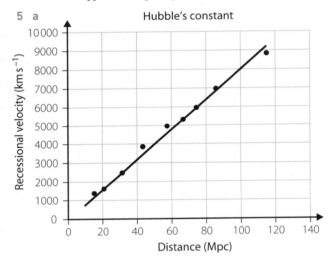

Hubble's constant

From the graph, Hubble's constant = gradient
$= 80.295 \, \text{km s}^{-1} \, \text{Mpc}^{-1}$.

b $\text{gradient} = \dfrac{80.295 \, \text{km}}{\text{Mpc s}} = \dfrac{80\,295 \, \text{m}}{3.085 \times 10^{22} \, \text{m s}}$

$\text{time} = \dfrac{3.085 \times 10^{22} \, \text{s}}{80\,295} = 3.842 \times 10^{17} \, \text{s}$

6   Although the precision of the measurements cannot be determined, the accuracy of the determination is limited given the calculated value falls outside the range of accepted values, being $73.8 \pm 2.4 \, \text{km s}^{-1} \, \text{Mpc}^{-1}$.

The reliability of the determination is very questionable because there is no evidence of repeated measurements of data points.

The number of data points used and the correlation shown by the line of best fit shows that a clear correlation has been achieved between the two data sets, supporting the idea of a valid conclusion.

WS 12.3 PAGE 110

1   Continuous spectra consist of all wavelengths produced by a black body emitter. Emission spectra consist of a number of distinct coloured lines corresponding to specific wavelengths emitted by electron transitions within a given substance. Absorption spectra are similar to continuous spectra, with distinct black lines on a continuous spectrum background where light has been absorbed by specific electron transitions in a relatively cool gas.

2   The lines present in an emission spectrum are produced by the emission of light from specific electronic transitions with an atom, which are unique to an individual element. An emission spectrum can be compared to a known standard to deduce the composition of the emitter because electron transitions, and thus emission lines, will be identical for the identical elements.

3   The distance to the star can be used to determine its recessional velocity. From this its red shift can be determined, and the spectrum can be shifted back by the same amount to create an accurate representation of its composition.

4   a   Mercury and sodium

    b   The 720 nm line and the 440 nm quartet

    c   Hydrogen

5   Spectra are produced as a result of electron transitions to different orbitals. Heavier elements contain more electrons to create transitions with a greater range of energies, thus producing more complex spectra.

6   Spectra gathered on Earth will have additional absorption lines due to light passing through Earth's atmosphere. This needs to be accounted for when determining chemical compositions.

WS 12.4 PAGE 112

1   The cooler outer layers of stars absorb photons of characteristic frequencies and produce absorption spectra.

2   Hotter stars will emit higher energy radiation. The energy of radiation is dependent on frequency according to $E = hf$.

This means that hotter stars must emit energy of higher

frequency, given that $\lambda = \dfrac{c}{f}$. This translates to a shorter

wavelength, making the star appear more blue.

3   Two of the factors affecting spectral line width are rotational speed and density. Higher density stars produce broader spectra but do not produce well defined edges in the lines. Rotational velocity also impacts the width of a spectral line: the faster the rotation, the greater the Doppler shifting at the edges and therefore the wider the spectral line. This leads to well-defined edges of a spectral line.

4   Photons are emitted and absorbed multiple times within the outer layers of a star. The greater the density of the star, the greater the number of absorptions and emission that will occur. As each absorption and emission occurs between atoms moving at high speeds, the wavelength of the photon emitted or absorbed is varied. This leads to a wider range of wavelengths about a normal emission line.

5 a A very hot blue star with significant luminosity, likely a blue giant or blue Main Sequence star.

b A cool red star with very low luminosity, most likely a red dwarf Main Sequence star; small likelihood of a red giant with relatively high luminosity.

c A relatively cool orange-red star with relatively low to relatively high luminosity; most likely a Main Sequence red dwarf but may be a Post-Main Sequence red giant.

6 a F class star

b B class star

c M class star

d G class star

WS 12.5 PAGE 114

1

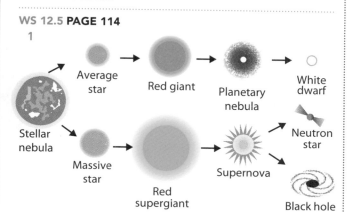

2 a Green peak wavelength, hydrogen absorption lines

b Blue peak wavelength, strong UV emission

c Red peak wavelength, strong IR emissions

3 a Red giant or red supergiant

b Blue main sequence or blue giant

4 Luminosity is an absolute measure of radiated electromagnetic power that depends on both the surface area and surface temperature. Red giants can have enormous surface area, so despite their lower surface temperature they can have a high luminosity.

5

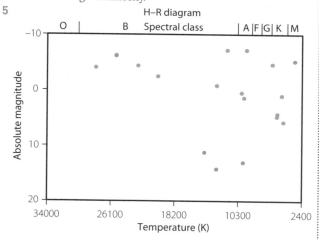

Depending on accuracy of plotting and interpretation of group boundaries, allocations of star group types may vary from answers, particularly distinguishing between Main Sequence stars and Red Giants towards top right corner of the graph.

Star name	Spectral class	Star group type
Sun	G	Main Sequence
Alpha Centauri A	G	Main Sequence
Achemar	B	Main Sequence
Sirius A	A	Main Sequence
Vega	A	Main Sequence
Rigel	B	Main Sequence
Hadar	B	Main Sequence
Deneb	A	Main sequence
Regulus	B	Main Sequence
Bellatrix	B	Main Sequence
Alpha Centauri B	K	Main Sequence
Antares	M	Red giant
Pollux	K	Main Sequence
Acrux	B	Main Sequence
Beta Centauri	B	Main Sequence
Polaris	G	Main Sequence
Sirius B	B	White dwarf
Procyon B	A	White dwarf
van Maanen's Star	B	White dwarf

WS 12.6 PAGE 116

1 Energy $= mc^2 = (5.0 \times 10^3)(3.00 \times 10^8)^2 = 4.5 \times 10^{20}$ J

2 Mass defect
$= 1.675 \times 10^{-27} - (1.673 \times 10^{-27} + 9.109 \times 10^{-31})$
$= 1.089 \times 10^{-30}$ kg
Energy released $= (1.089 \times 10^{-30})(3.00 \times 10^8)^2$
$= 9.801 \times 10^{-14}$ J

3 a Mass defect
$= 4(1.673 \times 10^{-27}) - (6.646 \times 10^{-27} + 2(9.109 \times 10^{-31}))$
$= 4.417\,82 \times 10^{-29}$ kg
Energy in joules $= (4.417\,82 \times 10^{-29})(3.00 \times 10^8)^2$
$= 3.976\,038 \times 10^{-12}$ J
Energy per nucleon $= \dfrac{3.976\,038 \times 10^{-12}}{4}$
$= 9.940\,095 \times 10^{-13}$
$= 9.94 \times 10^{-13}$ J

b Energy in joules $= (7.900 \times 10^{-28})(3.00 \times 10^8)^2$
$= 7.11 \times 10^{-11}$ J
Energy per nucleon $= \dfrac{7.11 \times 10^{-11}}{236}$
$= 3.012\,711\,864 \times 10^{-13}$
$= 3.01 \times 10^{-13}$ J

c Nuclear fusion of hydrogen produces more energy per nucleon than nuclear fission of uranium.

4 Kinetic energies of the product particles and electromagnetic radiation in the form of gamma rays.

5 Chemical reactions have a significantly smaller mass defect per nucleon and hence the energy released is much smaller. Nuclear reactions have a higher mass defect per nucleon and therefore release more energy.

WS 12.7 **PAGE 117**

1 The p–p chain is a rudimentary reaction requiring comparatively low temperatures and pressures. It requires only protons but is very slow due to the first step taking, on average, 9 billion years to occur. The CNO cycle uses carbon, which is present only in larger stars, to accelerate the production of helium. It requires higher temperatures and pressures but is much more efficient in producing helium.

2 The CNO cycle requires the fusion of a proton to a carbon nucleus. There are significantly higher coulomb repulsive forces to overcome than in the p–p chain and therefore it needs higher temperature and pressure.

3 While the p–p chain and the CNO cycle are vastly different processes, they have the same net reaction – carbon acts only as a catalyst. The same net reaction will release the same net energy because it produces the same mass defect.

4 $\dfrac{4.5 \times 10^{20}}{4.14 \times 10^{-12}} = 1.086\,956\,5 \times 10^{32} = 1.1 \times 10^{32}$ nuclei

5 An O class star would be almost exclusively powered by the CNO cycle, an A class star would have a significant amount of each of these reactions whereas the Sun would be almost exclusively powered by the p–p chain.

Chapter 13: Structure of the atom

WS 13.1 **PAGE 118**

1 Cathode rays travel in straight lines – as demonstrated by the shadow produced by a Maltese cross placed in a cathode ray tube (CRT).
Cathode rays are deflected by an electric field – as demonstrated by applying an electric field perpendicular to the cathode rays.
Cathode rays are deflected by a magnetic field – as demonstrated by applying a magnetic field perpendicular to the cathode rays.
Cathode rays have momentum – as demonstrated by the movement of a mica paddle wheel placed in the path of the cathode rays.
Cathode rays cause fluorescence – as demonstrated by the green glow produced against the glass of a CRT and by the production of light from a fluorescent material. This was later identified as a result of radiometric effects.
Cathode rays are identical regardless of their source.

2 At the time there was great debate as to whether cathode rays were waves or particles. Deflection by electric and magnetic fields provided strong evidence for cathode rays being negatively charged particles, as waves are not affected by magnetic or electric fields. The paddle wheel experiment provided further evidence for cathode rays

being a stream of particles, because the transfer of momentum was considered (at the time) to only be a property of particles.

3 That electrons were particles that could experience centripetal force.

4 The charge-to-mass ratio of an electron was extremely high (order of 10^{12}), which means that either the electron holds enormous charge or its mass is exceedingly small.

5 Sample answer

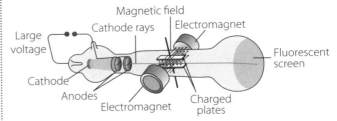

6 The anode was designed to collimate the beam of cathode rays, ensuring that all rays had the same initial velocity.

7 First we need to convert to SI units. $5\,\text{mm} = 0.005\,\text{m}$
Then substitute in $V = Ed = (80\,000)(0.005) = 400\,\text{V}$

8 Given $v = \dfrac{E}{B}, \dfrac{80\,000}{0.45} = 177\,777.7778\,\text{m s}^{-1} = 2 \times 10^5\,\text{m s}^{-1}$

9 $\dfrac{q}{m} = \dfrac{v}{rB}$ Rearranging gives us $r = \dfrac{vm}{Bq}$.

Substituting in, we get $r = \dfrac{(2.1903 \times 10^8)(9.109 \times 10^{-31})}{(7.419 \times 10^{-2})(1.602 \times 10^{-19})}$

$= 0.000\,168\,077\,25\,\text{m} = 1.681 \times 10^{-4}\,\text{m}$

10 Millikan's results indicated that charge was quantised. This meant that an atom must contain discrete amounts of both positive and negative charge, rather than a continuous range.

11 The top plate would have to be positively charged with the bottom plate negatively charged in order to attract the negatively charged droplets upwards against the downward force of gravity.

12 This would imply that the oil droplet has picked up four electrons or has a net electric charge equivalent to four electrons as a result of its ionisation.

13 Given $m = \dfrac{qE}{g}, \dfrac{(-1.6 \times 10^{-19})(65\,000)}{-9.8} = 1.061\,22 \times 10^{-15}\,\text{kg}$
$= 1.1 \times 10^{-15}\,\text{kg}$

14 First, we need to convert to SI units. $55\,\text{mm} = 0.055\,\text{m}$

Given $V = \dfrac{mgd}{q}, \dfrac{(4.6275 \times 10^{-16})(-9.8)(0.055)}{1.92 \times 10^{-18}} = -129.907\,\text{V}$
$= -1.3 \times 10^2\,\text{V}$

15 This experiment could be modelled with the use of a number of containers (20 or more) representing the oil droplets, each filled with various numbers of uniform objects (such as a single denomination coin) representing the charge on the oil droplet, and an electronic balance. Each of the containers would be weighed and plotted, the difference in mass between subsequent containers could be plotted on a frequency histogram with the smallest difference being the 'fundamental mass' of the uniform objects, representing the fundamental charge on an electron.

16 Models provide a simplified representation of a physics phenomenon that is often much more accessible then the 'real thing'. They are, however, limited in their complexity and accuracy and therefore will have limited predictive capabilities. Due to their simplicity and accessibility, models play a significant positive role for physics understanding and education.

WS 13.2 PAGE 122

1 Rutherford's model incorporated the electrons in a single orbit outside the small, dense, positive nucleus. Between the nucleus and electron orbit was a large amount of empty space.

2 Rutherford's model could not explain why the negatively charged electrons did not emit electromagnetic radiation due to centripetal acceleration and therefore collapse into the positively charged nucleus as a result of electrostatic attraction.

3 a The alpha particle passes through the empty space between the positive nucleus and the orbiting electrons.

   b The alpha particle collided with a positive nucleus with an angle of 45° to the tangent of the nucleus.

   c The alpha particle collided with a positive nucleus head-on, causing it to reflect back.

4 Thomson's model suggested a solid atom with low density but this could not explain the deflection of alpha particles at large angles.

5 They are both positively charged and therefore repel each other.

6 $^4_2\text{He} + ^9_4\text{Be} \rightarrow ^{12}_6\text{C} + ^1_0\text{n}$

7 The nucleus was updated to contain both positive and neutral particles rather than a single large positive sphere. Also, Rutherford had proposed that the atomic mass was proportional to the nuclear charge, but he could not reconcile this idea with the mass of hydrogen. If all nuclei other than hydrogen contained uncharged particles with a mass very close to that of the proton, the discrepancy could be explained.

8 When this radiation was incident upon paraffin wax, it ejected protons from the wax. Considering the conservation of momentum, the energy required to provide enough momentum to the protons was higher than had been measured for any form of radiation at that time, and therefore was more likely to be provided by a particle with similar mass to the proton that could transfer energy and momentum upon collision with the proton.

9 The electron is both electrically charged and easily removed from an atom, making it easy to isolate. The neutron, on the other hand, is not electrically charged and therefore unaffected by electric and magnetic fields. It is also strongly bonded to the nucleus, making it difficult to isolate. In addition, an electron is a stable particle outside the atom whereas free neutrons will decay, limiting the time frame in which they can be observed.

10 Scientific models are constantly refined over time as further evidence is generated from experimental results. The atomic model is an excellent example of this, as it developed over time from Thomson's plum pudding model, proposed as a result of the discovery of the electron, to Rutherford's adapted model suggesting, on the basis of the Geiger-Marsden experiment, that most of the atom was empty space. Rutherford's model was brought forward by the discovery of the neutron. The atomic model will continue to develop as more evidence is presented regarding its structure, make-up and interactions. Scientific models are constantly evolving to better approximate real-world phenomena.

Chapter 14: Quantum mechanical nature of the atom

WS 14.1 PAGE 125

1 Rutherford's model was unable to explain the stability of orbits, nor could it explain the lack of emitted EMR due to the acceleration of electrons in their circular orbits.

2 Bohr postulated that:
   1 Electrons exist in stable circular orbits
   2 Electrons in stable orbits do not emit or absorb radiation
   3 Electrons that transition between these stable orbits will emit or absorb specific amounts of energy corresponding to the energy difference between orbits.
   4 The angular momentum of electrons in orbit about the nucleus is quantised.

3 Bohr's model was able to explain the emission spectrum of hydrogen as being due to transitions of electrons between orbits. It was also able to explain the stability of atoms due to the quantum nature of energy emission.

4 Bohr contributed to what is now the quantum mechanical understanding of the atom, a major deviation from the classical mechanics that Rutherford used in an attempt to explain his model. The quantisation of electron orbits and the forbidden 'space' between them is something that cannot be explained by classical physics.

5 a Multi-electron atoms have variations in their spectral lines due to electron–electron interactions. These interactions make some transitions more energetic and some less energetic. Bohr's model considered only a single electron in isolation.

   b The Zeeman effect is the splitting of individual spectral lines when a magnetic field is applied to the excited atoms upon emission. The split is very small and difficult to resolve.

   c It could not explain the relative intensity of spectral lines – the observations that some spectral lines are much more intense than others.

   It could not explain the hyperfine structure of spectral lines – the splitting of spectral lines due to interactions with the magnetic moment of the nucleus.

WS 14.2 PAGE 127

1 Using $E = hf$, we see that $E \propto f$ and, given that $\lambda = \dfrac{c}{f}$, we see that shorter wavelength light (such as blue) will have a higher energy per photon than longer wavelength light (such as red).

2 a $f = \dfrac{c}{\lambda} = \dfrac{3.00 \times 10^8}{6.56 \times 10^{-7}} = 4.573\,17 \times 10^{14}\,\text{Hz}$

   $E = hf = (6.626 \times 10^{-34})(4.573\,17 \times 10^{14})$
   $= 3.030\,18 \times 10^{-19} = 3.03 \times 10^{-19}\,\text{J}$

b $E = \dfrac{hc}{\lambda} = \dfrac{(6.626 \times 10^{-34})(3.00 \times 10^8)}{4.86 \times 10^{-7}}$

$= 4.090\,123\,457 \times 10^{-19} = 4.09 \times 10^{-19}\,\text{J}$

c $E = \dfrac{hc}{\lambda} = \dfrac{(6.626 \times 10^{-34})(3.00 \times 10^8)}{4.34 \times 10^{-7}}$

$= 4.580\,184\,433\,2 \times 10^{-19} = 4.58 \times 10^{-19}\,\text{J}$

d $E = \dfrac{hc}{\lambda} = \dfrac{(6.626 \times 10^{-34})(3.00 \times 10^8)}{4.10 \times 10^{-7}}$

$= 4.848\,292\,683 \times 10^{-19} = 4.85 \times 10^{-19}\,\text{J}$

3 According to classical physics, the negative electrons should emit light as they progressively decay towards the positive nucleus. This light would vary continuously depending on the energy of the electron.

4 Explaining spectra of simple elements such as hydrogen unlocked the intricacies of spectra for all elements. This allowed astronomers to determine the chemical composition of stars. Additionally, by comparing the stellar spectra to those produced on Earth, features such as translational and rotational velocity could be determined as well as density.

5 Lyman series is the transitions that end in the $n = 1$ shell.

For the first transition, $\dfrac{1}{\lambda} = R_{\text{H}}\left(\dfrac{1}{n_{\text{f}}^2} - \dfrac{1}{n_{\text{i}}^2}\right)$

$= 1.097 \times 10^7\left(\dfrac{1}{1^2} - \dfrac{1}{2^2}\right) = 8\,227\,500$

$\therefore \lambda = 1.215\,436\,038 \times 10^{-7} = 1.215 \times 10^{-7}\,\text{m}$

For the second transition, $\dfrac{1}{\lambda} = R_{\text{H}}\left(\dfrac{1}{n_{\text{f}}^2} - \dfrac{1}{n_{\text{i}}^2}\right)$

$= 1.097 \times 10^7\left(\dfrac{1}{1^2} - \dfrac{1}{3^2}\right) = 9\,751\,111.111$

$\therefore \lambda = 1.025\,524\,157 \times 10^{-7} = 1.026 \times 10^{-7}\,\text{m}$

For the third transition, $\dfrac{1}{\lambda} = R_{\text{H}}\left(\dfrac{1}{n_{\text{f}}^2} - \dfrac{1}{n_{\text{i}}^2}\right)$

$= 1.097 \times 10^7\left(\dfrac{1}{1^2} - \dfrac{1}{4^2}\right) = 10\,284\,375$

$\therefore \lambda = 9.723\,48\,8301 \times 10^{-8} = 9.723 \times 10^{-8}\,\text{m}$

For the fourth transition, $\dfrac{1}{\lambda} = R_{\text{H}}\left(\dfrac{1}{n_{\text{f}}^2} - \dfrac{1}{n_{\text{i}}^2}\right)$

$= 1.097 \times 10^7\left(\dfrac{1}{1^2} - \dfrac{1}{5^2}\right) = 10\,531\,200$

$\therefore \lambda = 9.495\,594\,044 \times 10^{-8} = 9.496 \times 10^{-8}\,\text{m}$

For the fifth transition, $\dfrac{1}{\lambda} = R_{\text{H}}\left(\dfrac{1}{n_{\text{f}}^2} - \dfrac{1}{n_{\text{i}}^2}\right)$

$= 1.097 \times 10^7\left(\dfrac{1}{1^2} - \dfrac{1}{6^2}\right) = 10\,665\,277.78$

$\therefore \lambda = 9.376\,220\,862 \times 10^{-8} = 9.376 \times 10^{-8}\,\text{m}$

6 a $\lambda = \dfrac{c}{f} = \dfrac{3.00 \times 10^8}{2.459 \times 10^{15}} = 1.22 \times 10^{-7}\,\text{m}$. As can be seen from question **5**, a transition from $n = 2$ to $n = 1$ will produce a photon of wavelength $122\,\text{nm}$. It is likely then that the absorption of this wavelength will be due to a transition from $n = 1$ to $n = 2$.

b Rearranging $\dfrac{1}{\lambda} = R_{\text{H}}\left(\dfrac{1}{n_{\text{f}}^2} - \dfrac{1}{n_{\text{i}}^2}\right)$ to solve for n_{i} we get

$\dfrac{1}{n_{\text{i}}^2} = \dfrac{1}{n_{\text{f}}^2} - \dfrac{1}{\lambda R_{\text{H}}} = \dfrac{1}{1^2} - \dfrac{1}{(97 \times 10^{-9})(1.097 \times 10^7)} = 0.06$

$\therefore n_{\text{i}}^2 = 16.6$ so $n_{\text{i}} = 4$

The electron is initially in shell 4.

7 The Rydberg equation would need to be adjusted for other elements because, with an increase in the number of protons in the nucleus, the electrostatic force between nucleus and electrons increases. The electrons also encounter electrostatic repulsion among them and this would also need to be accounted for by using an R value relevant to each element.

WS 14.3 PAGE 130

1 a Detectors in the human eye do not have the sensitivity to detect individual wavelengths. Rather, the eye combines wavelengths to form colour images in the brain.

b Spectroscopes use a prism or diffraction grating to spread out the light, based on wavelength. The spread-out light is then observed against a wavelength scale.

2 Hydrogen has only four transitions in the visible region. The remaining transitions are in the ultraviolet region of the spectrum.

3 It can be concluded that the energy difference between $n = 1$ and $n = 2$ is the largest between adjacent energy levels because transitions to $n = 1$ are exclusively in the ultraviolet and photon energy is inversely proportional to wavelength. Transitions to $n = 3$ are exclusively in the infrared region. This suggests that the energy gap between subsequent orbits is smaller than the previous one.

4 Conservation of energy suggest that all of the energy absorbed during an excitation must be equal to the energy released during a relaxation, making the frequency of light produced the same.

5 a Risks could include high voltage used to stimulate the emission or high temperature associated with this process. Precautions would be to maintain distance between source and observer, ensure power is turned off before manipulating equipment and allowing sufficient time for equipment to cool before manipulation.

b Random errors may include parallax error in determining the position of the spectral line against a scale: the graduations on a classroom spectroscope are typically 20 or 50 nm, leading to significant opportunity for parallax error. Systematic errors might include an improperly calibrated scale.

WS 14.4 PAGE 132

1 The large momentum associated with large objects such as humans would yield a wavelength less than the Planck length, meaning no modern physics equipment is able to measure it. The multiple wavefronts associated with a macroscopic object are incoherent, further complicating the observations of these waves.

2 Considering electrons as waves, and thus existing as standing waves around the nucleus, implied that only certain energies would be stable (those energies that could produce a standing wave of fixed wavelength). It also supported the idea that these orbits did not decay into the nucleus.

3 Diffraction patterns are a property unique to waves. Electrons, and later neutrons, provided diffraction patterns of the crystalline structure of materials, similar to those produced by X-rays. Further to this, Davison and Germer were able to measure the wavelength of an electron by the diffraction pattern produced by an annealed nickel crystal. These determinations matched those of de Broglie's predictions.

9780170449687

4 a $\lambda = \dfrac{h}{mv} = \dfrac{6.626 \times 10^{-34}}{(9.109 \times 10^{-31})(2.52 \times 10^6)}$

$= 2.886557338 \times 10^{-10} = 2.89 \times 10^{-10}\,\text{m}$

b $\lambda = \dfrac{h}{mv} = \dfrac{6.626 \times 10^{-34}}{(1.675 \times 10^{-27})(2.52 \times 10^6)}$

$= 1.569770197 \times 10^{-13} = 1.57 \times 10^{-13}\,\text{m}$

c $\lambda = \dfrac{h}{mv} = \dfrac{6.626 \times 10^{-34}}{(6.64477 \times 10^{-27})(2.52 \times 10^6)}$

$= 3.957044532 \times 10^{-14} = 3.96 \times 10^{-14}\,\text{m}$

5 Visible light has wavelengths between 400 and 740 nm but the wavelength of an electron can be as small as

$\dfrac{h}{mv} = \dfrac{6.626 \times 10^{-34}}{(9.109 \times 10^{-31})(2.99 \times 10^6)} = 0.002\,\text{nm}$. This smaller

wavelength provides a much finer diffraction pattern, producing a higher resolution image.

WS 14.5 **PAGE 134**

1 An orbital is a region within which the probability of finding an electron exists. Orbitals have a variety of shapes and sizes depending on the orbital angular momentum of the electrons that occupy it.

2 Bohr's orbits were simple discrete concentric spheres with no set limit on the number of electrons they could contain, whereas Schrödinger's orbitals are spatial regions of probability limited to housing two electrons. Schrödinger's orbitals could also be repeated at higher energies.

3 Schrödinger developed a wave equation that could be used to explain the motion of electrons as waves rather than particles. The various solutions gave a probability of finding an electron at any given position rather than an exact path. This gave rise to the atomic orbital theory, which went on to explain not only the stability of an atom but chemical bonding as well.

4

Electron cloud

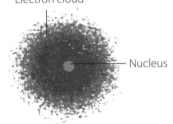

Nucleus

5 a 'Schrödinger's cat' is a thought experiment that describes a cat left inside a box with a vial of cyanide that would be broken (releasing the poison) when a detector detects the emission of a decay process that has a 50% chance of occurring while the cat is in the box. Unless the box is opened and the cat is observed, it can be thought of as both alive and dead simultaneously.

b Due to an electron exhibiting both wave and particle natures, it can be considered to be fulfilling all states of existence until observed in a particular state. This proposition is known as the Copenhagen interpretation.

c In the thought experiment, the cat can be considered as both dead and alive simultaneously. However, a cat is an extremely complex organism compared to a particle such as an electron, and as such cannot be represented in such a way.

Chapter 15: Properties of the nucleus

WS 15.1 **PAGE 136**

1 Alpha decay occurs when the number of protons is too high ($Z > 82$).

Beta decay occurs when the proton:neutron ratio is unstable. β^- decay occurs when there are too many neutrons, β^+ decay occurs when there are too few neutrons.

Gamma decay occurs when the nucleus is in an excited state.

2 Alpha decay removes two neutrons and two protons from the parent nucleus. This can create instabilities in the proton-to-neutron ratio of heavy nuclei and hence beta decay often occurs in combination with alpha decay to achieve stability.

When a nucleus decays through the ejection of a particle, it is often left in an excited state and subsequently undergoes gamma decay in order to achieve its ground state.

3 $^{229}_{90}\text{Th} \rightarrow \,^{225}_{88}\text{Ra} + \,^{4}_{2}\text{He}$

4 $^{14}_{6}\text{C} \rightarrow \,^{14}_{7}\text{N} + \,^{0}_{-1}\text{e}$

5 In general, the greater the Z number is (above $Z = 82$) the less stable the nucleus is. This means that elements with extremely high Z numbers have half-lives in the order of nanoseconds, making them very difficult to observe and identify.

6 a $^{99}_{43}\text{Tc} \rightarrow \,^{99}_{44}\text{Ru} + \,^{0}_{-1}\text{e}$

b $^{23}_{12}\text{Mg} \rightarrow \,^{23}_{11}\text{Na} + \,^{0}_{1}\text{e}$

c $^{219}_{86}\text{Rn} \rightarrow \,^{215}_{84}\text{Po} + \,^{4}_{2}\text{He}$

7 a $^{228}_{90}\text{Th} \rightarrow \,^{224}_{88}\text{Ra} + \,^{4}_{2}\text{He}$

b $^{131}_{53}\text{I} \rightarrow \,^{131}_{54}\text{Xe} + \,^{0}_{-1}\text{e}$

c $^{11}_{6}\text{C} \rightarrow \,^{11}_{5}\text{B} + \,^{0}_{1}\text{e}$

WS 15.2 **PAGE 138**

1 36.0348 hours is exactly 6 half-lives. Therefore the mass

would be equal to $\dfrac{345}{2^6} = 5.3906\,\text{g} = 5.39 \times 10^{-3}\,\text{kg}$. (Note that

this could have been answered using the formulae $N_t = N_0 e^{-\lambda t}$

and $\lambda = \dfrac{\ln 2}{t_{1/2}}$ but that would take much longer!)

2 First, we need to convert 5.27 years into seconds.

$t_{1/2} = (5.27)(365.25)(24)(60)(60) = 166\,308\,552\,\text{s}$

$\lambda = \dfrac{\ln 2}{t_{1/2}} = \dfrac{\ln 2}{166\,308\,552} = 4.167838 \times 10^{-9} = 4.2 \times 10^{-9}\,\text{s}^{-1}$

3 a $t_{1/2} = \dfrac{\ln 2}{\lambda} = \dfrac{\ln 2}{2.1 \times 10^{-6}} = 330\,070 = 3.3 \times 10^5\,\text{s}$

b Rn-222 is more active than Co-60, given its larger decay constant (500 times larger) and shorter half-life (by the same factor of 500).

4 Converting 2.65 years to seconds,

$t = (2.65)(365.25)(24)(60)(60) = 83\,627\,640\,\text{s}$

$N_t = N_0 e^{-\lambda t} = 6.022 \times 10^{23}\, e^{-(5.81 \times 10^{-8})(83\,627\,640)} = 4.6731 \times 10^{21}$

$= 4.7 \times 10^{21}\,\text{nuclei}$

5 $N_t = N_0 e^{-\lambda t}$. Rearranging to solve for λ:

$\lambda = \dfrac{\ln\left(\dfrac{N_t}{N_0}\right)}{-t} = \dfrac{\ln\left(\dfrac{0.72045}{1}\right)}{-(46.21)(60)(60)}$

$= 1.970949 \times 10^{-6} = 1.971 \times 10^{-6}\,\text{s}^{-1}$

6 $N_t = N_0 e^{-\lambda t}$. Rearranging to solve for λ:

$$\lambda = \frac{\ln\left(\dfrac{N_t}{N_0}\right)}{-t} = \frac{\ln\left(\dfrac{1\,499\,998\,625}{1\,500\,000\,000}\right)}{-(15.00)(365.25)(24)(60)(60)}$$

$$= 1.936\,495 \times 10^{-15}\,\text{s}^{-1}$$

$$t_{1/2} = \frac{\ln 2}{\lambda} = 3.57939 \times 10^{14} = 3.6 \times 10^{14}\,\text{s}$$

WS 15.3 **PAGE 140**

1 Uncontrolled reactions occur when the rate of reaction increases exponentially. For example, the fission of U-235 requires the capture of a single neutron but produces three additional neutrons. If each of these neutrons also initiates a fission reaction, that produces a further nine neutrons in total. Subsequent reactions can produce up to 27, 81, 243 ... neutrons.

2 Exactly one neutron would be required to absorbed by subsequent nuclei. This is achieved with the use of control rods. These rods are made of neutron-absorbing material and are inserted in between fuel rods to absorb neutrons. The depth to which control rods are inserted will determine the rate of reaction.

3 Most of the energy produced in a fission reaction is carried away as kinetic energy of the daughter nuclei.

4 If not all the neutrons released by fission are able to be captured, it means either that a suitable isotope must produce more than one neutron per reaction or that an additional external source of neutrons is required. The use of an external source is becoming more common in modern reactors.

5 The ping-pong balls represented the neutrons.

6 The mouse trap model is a very simple model that is easy to visualise. Both mouse traps and fissionable material are unstable. In the model, the ping-pong balls released initiate new reactions, similar to what happens in fission reactions. However, this reaction rate is much slower, because it releases only one ping-pong ball per reaction compared to three neutrons in the fission of U-235. Also, the ping-pong balls are capable of initiating multiple subsequent reactions, whereas in fission the neutrons are captured by a nucleus. The scaling of the model is very poor, because the nuclei in fission reactions are much further apart than the mouse trap model suggests, and the energies of the ping-pong balls do not vary like those of neutrons.

WS 15.4 **PAGE 142**

1 Spontaneous transmutation is a mutation that occurs without human interaction, whereas artificial transmutation occurs as a result of bombardment by human means.

Spontaneous transmutation: $^{238}_{92}\text{U} \rightarrow\ ^{234}_{90}\text{Th} +\ ^{4}_{2}\text{He}$

Artificial transmutation: $^{59}_{27}\text{Co} +\ ^{1}_{0}\text{n} \rightarrow\ ^{60}_{27}\text{Co}$

2 Nuclear isotopes that undergo artificial transmutation require bombardment by other particles, something that can be controlled. If sufficient mass of a spontaneous fissile material were used, it would not be able to be controlled and would produce a runaway reaction.

3 High temperatures are required to provide the nucleons with enough energy to overcome the electrostatic repulsion between the two positive nuclei approaching each other, enabling the nuclei to collide. The increased pressure increases the chance of a collision and hence the number of collisions so there are enough to sustain the thermonuclear fusion.

4 Nuclei above iron decrease in stability and, therefore, the binding energy per nucleon is also reduced. This means that the production of elements such as lead by fusion will

result in a net transfer of energy into the nucleus rather than energy being released from it.

5 As identified in question **4**, a net input of energy is required to produce nuclei heavier than iron. Although these reactions are typically only observed during the supernovae phase of Post-Main Sequence stars, they can also occur during other high energy events such as the collision of neutron stars or in particle accelerator collisions.

6 A single proton is not bound to any other nucleon and therefore has no binding energy.

7 If we take the example of fusing hydrogen to helium and the fission of uranium to its daughter nuclei, the difference in binding energy per nucleon between reactants and products is far greater for the fusion of hydrogen than for the fission of uranium. Thus it could be concluded that fusion releases more energy per nucleon than fission.

WS 15.5 **PAGE 144**

1 Mass defect $= (235.044 + 1.008) - (143.923 + 88.917 + 3(1.008))$
$= 0.188\,u\,(0.188)(1.661 \times 10^{-27}) = 3.122\,68 \times 10^{-28}$
$= 3.123 \times 10^{-28}\,\text{kg}$

2 Mass defect $= 2(1.673 + 1.675) \times 10^{-27} - (6.644\,77 \times 10^{-27})$
$= 0.069\,448 \times 10^{-27}\,\text{kg}$
$E = (6.9448 \times 10^{-29})(3.00 \times 10^8)^2 = 6.250\,32 \times 10^{-12}$
$= 6.25 \times 10^{-12}\,\text{J}$

3 $E = (5.00 \times 10^9)(3.00 \times 10^8)^2(60)(60)(24) = 3.888 \times 10^{31}$
$= 3.9 \times 10^{31}\,\text{J}$

4 Mass defect $= 14(1.673 + 1.675) \times 10^{-27} - (27.977)(1.661 \times 10^{-27})$
$= 0.402\,203 \times 10^{-27}\,\text{kg}$
$E = (4.022\,03 \times 10^{-28})(3.00 \times 10^8)^2 = 3.619\,827 \times 10^{-11}\,\text{J}$
$= \dfrac{3.619\,827 \times 10^{-11}}{1.602 \times 10^{-19}} = 2.259\,567\,416 \times 10^8 = 2.26 \times 10^8\,\text{eV}$

5 Mass defect $= (26)\left(\dfrac{1.673}{1.661}\right) + (30)\left(\dfrac{1.675}{1.661}\right) - 55.935$
$= 0.505\,698\,374\,5\,u$

$E = \dfrac{(0.505\,698\,374\,5)(3.00 \times 10^8)^2}{9.315 \times 10^8}$
$= 4.830\,917\,874 \times 10^7\,\text{eV}$
$= \dfrac{4.830\,917\,874 \times 10^7}{56}$
$= 8.626\,639\,061 \times 10^7 = 8.63 \times 10^7\,\text{eV per nucleon}$

6 a **Step 1:** alpha decay
mass defect $= 238.051 - (234.044 + 4.0026)$
$= 0.0044\,u$
$E = (0.0044)(931.5) = 4.0986 = 4.099\,\text{MeV}$
Converting to joules we get:
$(4.0968 \times 10^6)(1.602 \times 10^{-19}) = 6.565\,957\,2 \times 10^{-13}$
$= 6.566 \times 10^{-13}\,\text{J}$

Step 2: beta decay
mass defect $= 234.044 - (234.043 + 0.000\,548\,404\,57)$
$= 0.000\,451\,595\,u$
$E = (0.000\,451\,595)(931.5) = 0.420\,661 = 0.4207\,\text{MeV}$
Converting to joules we get:
$(0.420\,661\,143 \times 10^6)(1.602 \times 10^{-19}) = 2.166\,161\,509 \times 10^{-13}$
$= 2.166 \times 10^{-13}\,\text{J}$

Step 3: beta decay
mass defect $= 234.043 - (234.041 + 0.000\,548\,404\,57)$
$= 0.001\,451\,595\,u$
$E = (0.001\,451\,595\,43)(931.5) = 1.352\,160 = 1.352\,\text{MeV}$
Converting to joules we get:
$(1.352\,161\,143 \times 10^6)(1.602 \times 10^{-19}) = 6.738\,991\,512 \times 10^{-14}$
$= 6.739 \times 10^{-14}\,\text{J}$

9780170449687

Step 4: alpha decay

mass defect $= 234.041 - (230.033 + 4.0026)$
$$= 0.0054\,u$$

$E = (4.0054)(931.5) = 5.0301 = 5.030\,\text{MeV}$

Converting to joules we get:

$(5.0301 \times 10^6)(1.602 \times 10^{-19}) = 8.058\,22 \times 10^{-13} = 8.058 \times 10^{-13}\,\text{J}$

Step 5: alpha decay

mass defect $= 230.033 - (226.025 + 4.0026)$
$$= 0.0054\,u$$

$E = (4.0054)(931.5) = 5.0301 = 5.030\,\text{MeV}$

Converting to joules we get:

$(5.0301 \times 10^6)(1.602 \times 10^{-19}) = 8.058\,22 \times 10^{-13}$
$$= 8.058 \times 10^{-13}\,\text{J}$$

Step 6: alpha decay

mass defect $= 226.025 - (222.018 + 4.0026)$
$$= 0.0044\,u$$

$E = (0.0044)(931.5) = 4.0986 = 4.099\,\text{MeV}$

Converting to joules we get:

$(4.0986 \times 10^6)(1.602 \times 10^{-19}) = 6.565\,96 \times 10^{-13}$
$$= 6.566 \times 10^{-13}\,\text{J}$$

Step 7: alpha decay

mass defect $= 222.18 - (218.009 + 4.0026)$
$$= 0.0064\,u$$

$E = (0.0064)(931.5) = 5.9616 = 5.962\,\text{MeV}$

Converting to joules we get:

$(5.9616 \times 10^6)(1.602 \times 10^{-19}) = 9.550\,48 \times 10^{-13}$
$$= 9.550 \times 10^{-13}\,\text{J}$$

Step 8: alpha decay

mass defect $= 218.009 - (214.00 + 4.0026)$
$$= 0.0064\,u$$

$E = (0.0064)(931.5) = 5.9616 = 5.962\,\text{MeV}$

Converting to joules we get:

$(5.9616 \times 10^6)(1.602 \times 10^{-19}) = 9.550\,48 \times 10^{-13}$
$$= 9.550 \times 10^{-13}\,\text{J}$$

Step 9: beta decay

mass defect $= 214.00 - (213.999 + 0.000\,548\,404\,57)$
$$= 0.000\,451\,595\,u$$

$E = (0.000\,451\,595)(931.5) = 0.420\,661\,143 = 0.4207\,\text{MeV}$

Converting to joules we get:

$(0.420\,661\,143 \times 10^6)(1.602 \times 10^{-19}) = 6.738\,99 \times 10^{-14}$
$$= 6.739 \times 10^{-14}\,\text{J}$$

Step 10: beta decay

mass defect $= 213.999 - (213.995 + 0.000\,548\,404\,57)$
$$= 0.003\,451\,595\,43\,u$$

$E = (0.003\,451\,595\,43)(931.5) = 3.215\,161\,143 = 3.215\,\text{MeV}$

Converting to joules we get:

$(3.215\,161\,143 \times 10^6)(1.602 \times 10^{-19}) = 5.150\,69 \times 10^{-13}$
$$= 5.151 \times 10^{-13}\,\text{J}$$

Step 11: alpha decay

mass defect $= 213.995 - (209.984 + 4.0026)$
$$= 0.0084\,u$$

$E = (0.0084)(931.5) = 7.8246 = 7.825\,\text{MeV}$

Converting to joules we get:

$(7.8246 \times 10^6)(1.602 \times 10^{-19}) = 1.2535 \times 10^{-12} = 1.254 \times 10^{-12}\,\text{J}$

Step 12: beta decay

mass defect $= 209.984 - (209.983 + 0.000\,548\,404\,57)$
$$= 0.000\,451\,595\,43\,u$$

$E = (0.000\,451\,595\,43)(931.5) = 0.420\,661\,143 = 0.4207\,\text{MeV}$

Converting to joules we get:

$(0.420\,661\,143 \times 10^6)(1.602 \times 10^{-19}) = 6.738\,991\,511 \times 10^{-14}$
$$= 6.739 \times 10^{-14}\,\text{J}$$

Step 13: beta decay

mass defect $= 209.983 - (209.982 + 0.000\,548\,404\,57)$
$$= 0.000\,451\,595\,43\,u$$

$E = (0.000\,451\,595\,43)(931.5) = 0.420\,661\,143 = 0.4207\,\text{MeV}$

Converting to joules we get:

$(0.420\,661\,143 \times 10^6)(1.602 \times 10^{-19}) = 6.738\,991\,511 \times 10^{-14}$
$$= 6.739 \times 10^{-14}\,\text{J}$$

Step 14: alpha decay

mass defect $= 209.982 - (205.973 + 4.0026)$
$$= 0.0064\,u$$

$E = (0.0064)(931.5) = 5.9616 = 5.962\,\text{MeV}$

Converting to joules we get:

$(5.9616 \times 10^6)(1.602 \times 10^{-19}) = 9.550\,48 \times 10^{-13}$
$$= 9.550 \times 10^{-13}\,\text{J}$$

b The alpha decay steps typically release the most energy because the mass defect is larger in these steps.

Chapter 16: Deep inside the atom

WS 16.1 **PAGE 147**

1 It would suggest that other subatomic particles would also have antiparticles and that an entire system of antimatter could exist.

2 An electron has mass of $9.109 \times 10^{-31}\,\text{kg}$. Given a positron is identical to an electron except for charge, we can assume the combined mass is equal to $1.8218 \times 10^{-30}\,\text{kg}$. The total energy released according to $E = mc^2 = (1.8218 \times 10^{-30})(3.00 \times 10^8)^2 = 1.639\,62 \times 10^{-13}\,\text{J}$. If this is to be divided equally between two gamma rays then each would have $8.1981 \times 10^{-14}\,\text{J}$.

Using $\lambda = \dfrac{hc}{E}$ we get $\lambda = \dfrac{(6.626 \times 10^{-34})(3.00 \times 10^8)}{8.1981 \times 10^{-14}}$
$$= 2.4247 \times 10^{-12} = 2.42 \times 10^{-12}\,\text{m}.$$

3 Neutrons were involved in β decay, which was known to produce an electron (or a positron) and increase or decrease in the atomic number of the element, meaning a change in proton number.

4 β^- decay involves the transformation of a neutron into a proton and an electron. An example of this is the decay of carbon-14 according to the following equation:
$${}^{14}_{6}\text{C} \rightarrow {}^{14}_{7}\text{N} + {}^{0}_{-1}\text{e}.$$

5 β^+ decay involves the transformation of a proton into a neutron and a positron. An example of this is the decay of potassium-40 according to the following equation:
$${}^{40}_{19}\text{K} \rightarrow {}^{40}_{18}\text{Ar} + {}^{0}_{1}\text{e}.$$

6 It suggests that neither protons nor neutrons are fundamental particles because they can break down into more fundamental particles such as electrons and positrons.

7 Given the larger mass of a neutron, a decay of a free proton into a neutron would violate energy conservation laws so this cannot happen. (β^+ decay can occur within a nucleus through energy changes within the nucleus.)

8 Beta emission from the nucleus of unstable atoms such as C-14 indicated that the nucleus (or the particles within it) consisted of something other than just protons and neutrons.

Antimatter had been theorised and was confirmed upon the discovery of the positron. The suggestion that there could be antimatter particles for particles other than the electron added speculation that other subatomic particles existed.

Cosmic rays from the Sun interact with atoms in the upper atmosphere and produce a variety of detectable particles not seen before, including the muon.

The later discovery of neutrinos further supported theories that suggested or incorporated a wider group of subatomic and fundamental particles.

9 The discovery of various additional subatomic and fundamental particles led to the development of the Standard Model of matter, as well as changes to theories surrounding forces and the fields within which all matter exists.

WS 16.2 PAGE 150

1 A boson is any particle that has an integer value of spin. This includes the elementary bosons that mediate forces but also includes composite particles like the mesons.

2 The strong nuclear force acts as a strong attractive force between masses at the nucleon separation distance and is strongly repulsive at distances less than the nucleon separation distance. The strong nuclear force is responsible for holding nucleons together.

3 A subatomic particle is simply a particle that is smaller than an atom. Subatomic particles can be composite particles or indivisible particles. An example of a composite subatomic particle is the proton. In contrast, a fundamental particle is a particle that cannot be broken down into simpler particles. An example of a fundamental particle is the electron.

4 A baryon consists of an odd number of quark components (typically three). It can be a combination of either quarks or antiquarks. Baryons are also classified as fermions because they have half-integer spins. Mesons consist of a quark and anti-quark and are classed as bosons because they have integer spins.

5 Hadrons are particles composed of quarks. There are two types of hadrons: baryons (three-quark combinations) and mesons (two-quark combinations). A proton is a hadron made up of two up quarks and a single down quark.

6 Second and third generation quarks are very high energy particles and as such are inherently unstable. This leads to half-lives in the order of 1 billionth of a second and shorter. The top quark is so short-lived its half-life is estimated to be 10^{-25} s. Such short half-lives make it close to impossible to observe these particles directly. Typically, they are identified by their decay products and energies.

7 An antimatter hydrogen atom would be made up of one antiproton (two antiup quarks and one antidown quark) with an antielectron (or positron) orbiting it.

WS 16.3 PAGE 153

1 A neutron does not carry a net charge and, therefore, will be unaffected by an electric field. However, if it were to decay, its component quarks would be affected by both the magnetic and electric fields in a particle accelerator.

2 Basic kinematics tells us that $v = u + at$. Simply put, for a given acceleration (provided by the electric fields), the longer the electric field is applied, the greater the final velocity achieved. Synchrotrons accelerate their particles for much greater lengths of time than do cyclotrons or linear particle accelerators.

3 Using the special relativity relationship $p = \dfrac{mv}{\sqrt{1 - \dfrac{v^2}{c^2}}}$ and

substituting, we get $p = \dfrac{(1.673 \times 10^{-27})(299\,999\,970)}{\sqrt{1 - \left(\dfrac{0.999\,999\,9^2}{1^2}\right)}}$

$$= 1.22\,282\,434 \times 10^{-15}$$
$$= 1.122 \times 10^{-15}\,\text{kg m s}^{-1}$$

4 Evacuation of the accelerator is required for two main reasons. From a mechanical point of view, the particles would quickly lose momentum if they underwent frequent collisions. From a particle physics point of view, any collisions within the accelerator (other than the desired

ones) could initiate transmutations and decays in unwanted locations within the accelerator.

5 Using the special relativity relationship for momentum, as $v \rightarrow c$ the momentum of the particle $\rightarrow \infty$. This implies that in order for a particle to have infinite momentum, infinite force would have to be applied to the particle. Over any given time this would imply infinite impulse. This violates the conservation of momentum law.

6 As the particle accelerates it covers a greater distance in a given time. To maintain uniform acceleration, each tube must be longer than the previous tube to ensure the particle spends the same time in each tube.

7 Combining $F = Eq$ and $E = \dfrac{V}{d}$, we can rearrange to get $V = \dfrac{Fd}{q}$. Substituting, we get:

$$V = \frac{(7.1 \times 10^{-14})(1.4 \times 10^{-2})}{1.602 \times 10^{-19}} = 6204.744\,07 = 6200\,\text{V}$$

MODULE EIGHT: CHECKING UNDERSTANDING PAGE 156

1 **D** 4.22 million tonnes per second

2 **D** Electrons can absorb or release energy when they change levels.

3 **A** it quantified the charge on an electron.

4 **C** Beta negative

5 **B** Blue dwarf

6 Alpha decay typically occurs when the mass of a nucleus is too large for it to be stable ($Z > 82$). Beta negative decay occurs when proton-to-neutron ratio is too low. Beta positive decay occurs when the proton-to-neutron ratio is too high. Gamma decay occurs when the nucleus is in an excited state.

7 Cepheid variables are stars whose diameter and luminosity change rapidly. The period and magnitude of these changes are both well-defined and consistent. This allows scientists to determine the absolute magnitude of these stars and, by using the ratio of absolute and apparent magnitude, the distance to such stars can be calculated. This provides measurements of distances well beyond the limits of the Milky Way. Distances to cepheid variables was used in conjunction with the red shift to determine recessional velocity, which could then be plotted to determine the expansion rate of the Universe.

8 Every element produces a unique emission and adsorption spectrum. The lines found in stellar spectra can be matched to those of known elements to determine chemical composition.

9 In both processes, hydrogen is fused to produce helium and approximately 26.7 MeV of energy, but by different mechanisms. The p–p chain is present in smaller stars and is slower, because its first step is two protons fusing. The CNO cycle uses heavier isotopes to catalyse the reaction, with the first step being the fusion of a proton to a carbon nucleus. This process is much faster and occurs in higher mass stars.

10 Cathode rays travel in straight lines but are deflected by both magnetic and electric fields. They are capable of causing fluorescence and are perceived to have momentum.

11 By firing high energy alpha particles towards gold foil and measuring the angle at which the particles deflected, the experiment was able to conclude that most of an atom was empty space with a tiny, dense, positively charged nucleus placed at the centre.

12 1 Electrons exist in stable orbits.

    2 Electrons in stable orbits do not emit or absorb radiation.

    3 Electrons that transition between these stable orbits will emit or absorb specific amounts of energy corresponding to the energy difference between orbits.

    4 The angular momentum of electrons in orbit about the nucleus is quantised.

9780170449687

13 By combining $E = mc^2$ and $E = \dfrac{hc}{\lambda}$, wavelength can be represented as $\lambda = \dfrac{h}{mc}$. Substituting the velocity for the speed of light, this relationship could be used to determine the de Broglie wavelength of any object with momentum. The diffraction patterns of both electrons and neutrons supported the theory that particles can demonstrate wave behaviour.

14 Schrödinger's wave equation provides solutions for stable orbitals. These orbitals map out a region in which an electron could be found, rather than tracing a path that the electrons follow.

15 Cyclotrons are typically relatively small and produce lower energy collisions. They work by using an electric field to accelerate charged particles across a gap between D magnets, which direct the beam in circles before the beam is allowed to collide with a stationary target. Synchrotrons are much larger and produce much higher energy collisions. They also use electric fields to accelerate beams of charged particles and magnetic fields to collimate the beams in large doughnut tubes. They allow for collisions between opposing beams or with a stationary target.

16 A fermion is a particle with half-integer spin, such as an electron. Bosons, such as photons, have integer spin.

17 $^{238}_{92}U \rightarrow\ ^{234}_{90}Th + ^{4}_{2}He$

18 $\dfrac{1}{\lambda} = R_H\left(\dfrac{1}{n_f^{\,2}} - \dfrac{1}{n_i^{\,2}}\right) = 1.097 \times 10^7\left(\dfrac{1}{4^2} - \dfrac{1}{6^2}\right) = 380\,902.7778$

$\therefore \lambda = 2.625\,341\,84 \times 10^{-6} = 2.625 \times 10^{-6}\,m$

19 Convert 6 hours into seconds: $t_{1/2} = (6)(60)(60) = 21\,600$

$\lambda = \dfrac{\ln 2}{t_{1/2}} = \dfrac{\ln 2}{21\,600} = 3.209\,014\,725 \times 10^{-5} = 3 \times 10^{-5}$

20 Using $F_B = qvB$ and $F = \dfrac{mv^2}{r}$ we get $\dfrac{mv^2}{r} = qvB$.

Rearranging gives $r = \dfrac{mv}{qB}$.

Substituting, we get $r = \dfrac{(9.109 \times 10^{-31})(6.25 \times 10^6)}{(1.602 \times 10^{-19})(1.972 \times 10^{-4})}$

$= 1.802\,110 \times 10^{-4} = 1.80 \times 10^{-4}\,m$

Practice exam

SECTION 1 PAGE 159

1 **C** Because there are no horizontal forces acting on the projectile and there is a constant force down due to gravity, **C** is the correct answer. **A**, **B** and **D** all include horizontal forces.

2 **A** An electromagnetic wave with oscillating electric and magnetic fields.

The image clearly shows the induction of both electric and magnetic fields as in the classic electromagnetic wave diagram. It cannot be **B** or **C** because it does not show any particles moving and so it is not a mechanical wave. Option **D** suggests that electromagnetic waves require a medium to propagate through, which they do not.

3 **A** Young's double-slit experiment, due to the light and dark bands seen on the screen. These indicate an interference pattern as seen in wave behaviour.

The diagram shows the experimental set-up for the double-slit experiment carried out by Young. It also shows that you would see dark and light bands on the

screen, indicative of destructive and constructive wave behaviour. The photoelectric effect experiment does show that light of a certain frequency will cause electrons to leave a metal, but it does not demonstrate wave behaviour; nor does the diagram show this experiment, so it cannot be **B**. The rotating mirror experiment was used to measure the speed of light and not show the wave nature of light, so **C** is not the correct answer. **D** is correct in that Malus did carry out experiments to show the wave nature of light due to polarisation, but it does not use the experimental set-up shown.

4 **A** $\vec{F}_{by\,2\,on\,1} = -\vec{F}_{by\,1\,on\,2}$

Newton's Third Law states that for every force there must be an equal and opposite reaction force – the size of the currents and lengths of the wires will not affect this relationship. **B** is not correct because the absence of the minus sign suggests that the two forces act in the same direction, which is incorrect.

5 **A** $^{23}_{12}Mg \rightarrow\ ^{23}_{11}Na + ^{0}_{1}e + ^{0}_{0}\nu$

C and **D** feature beta-minus decay and are, therefore, incorrect. **B** is incorrect because it involves the wrong isotope of magnesium decaying. **A** is the only answer with no errors.

6 **C** As soon as there is no centripetal force acting on the object, its motion will be determined by Newton's First Law – an object will continue in its state of motion unless acted on by an unbalanced external force. Consequently, it will continue in the direction it was moving at the instant the centripetal force ceased, which is at a tangent to the circle at that point.

7 **A** $4.88 \times 10^{-6}\,m$

The maxima of the bands observed on the screen can be found using $m\lambda = d\sin\theta$. Rearranging we get $d = \dfrac{\lambda}{\sin\theta}$. Substituting in the values from the question gives $\dfrac{425 \times 10^{-9}}{\sin 5°} = 4.88 \times 10^{-6}\,m$.

8 **A** $7560\,m\,s^{-1}$

Earth's radius is $6.371 \times 10^6\,m$ (from the data sheet); therefore, the orbital radius is $7.001 \times 10^6\,m$. Earth's mass is $6.0 \times 10^{24}\,kg$ (data sheet). Substituting into the equation for orbital velocity $v = \sqrt{\dfrac{GM}{r}}$ yields a value of $7561\,m\,s^{-1}$.

9 **A** Wavelengths can be calculated using Rydberg's equation and then knowledge of the boundary wavelengths of the visible spectrum (400 nm violet to 700 nm red) can be used to determine which wavelengths belong in the infrared, visible and ultraviolet regions of the electromagnetic spectrum.

Alternatively, one could conclude that the electron falling from $n = 4$ to $n = 1$ will release the greatest amount of energy since it is the greatest electron shell difference. Since UV radiation is the most energetic form of radiation and has the shortest wavelength, then this electron 'descent' could be determined to correspond to UV and 97 nm without any calculations being required. Similarly, the 'descent' from $n = 4$ to $n = 3$ could be concluded to be the least energetic and therefore to correspond to infrared radiation and the longest wavelength.

10 **D** Using the right-hand rule it can be seen that the force on the wire will be perpendicular to the magnetic field and the current in the wire. In a radial field, this means that the direction of the force on the wire changes as the coil moves in the field in such a fashion that it is always

perpendicular to the pivot arm. Since torque is maximum when F and r are perpendicular, this results in an increase in overall torque as the coil rotates in the field.

11 **C** $3.52 \times 10^{-7}\,\text{m}$

As the wavelength increases it causes the energy of the photoelectron to decrease. So the largest wavelength to emit the photoelectron is the one that has just enough energy to cause it to be emitted. Using $E = hf$, we can substitute in $\frac{c}{\lambda}$ to give $E = \frac{hc}{\lambda}$. Rearranging to get λ we have $\lambda = \frac{hc}{E}$. Substituting in we have $\frac{(6.626 \times 10^{-34})(3.00 \times 10^{8})}{5.65 \times 10^{-19}} = 3.52 \times 10^{-7}\,\text{m}$.

12 **D** Baryons and mesons are both particles comprised of multiple quarks. Baryons are three-quark combinations and mesons are quark-antiquark combinations. All leptons exist alone, as do bosons. Consequently, the lambda must be a baryon, the pion a meson and the tau a lepton. The answer is, therefore, **D**.

13 **D** No transformer would be constructed with a single loop of wire in both the primary and secondary coil since that would be pointless (**C** is incorrect). Lenz's Law is a law because it always applies (**A** is incorrect). All real transformers feature a laminated core, yet none are ideal because all will transform some electrical energy into heat, so **B** is incorrect. The equation used to quantitatively analyse ideal transformers assumes no energy 'loss', which would require perfect flux linkage (**D**).

14 **B** Essential to answering this question is understanding the theory of Rutherford (based on the gold foil experiment) and Bohr (based on observations of emission spectra and Planck's quantisation of energy $E = hf$). As is usual for models, their models had shortcomings. Rutherford's model featured electrons moving in circles but according to classical physics the electrons would then be accelerating charged particles; they should emit energy and spiral into the nucleus. Bohr's model could not explain several observed spectral features such as hyperfine spectral lines, the Zeeman effect and the differences in intensities of spectral lines.

15 **B** In order to balance the solid plank, the clockwise torque and anticlockwise torque need to be of equal magnitude. The anticlockwise torque is given by the force due to gravity acting on the left child times the distance from the pivot ($31 \times 9.8 \times 3.0 = 911.4\,\text{N}$). Therefore the clockwise torque, given by the force due to gravity acting on the right child times the distance from the pivot) must be the same. Solving for $m \times 9.8 \times 4.0$ yields a value of $23.25\,\text{kg}$ (**B**). Alternatively, you could conclude as the right-hand child is sitting further from the pivot that child must have a lower mass, hence eliminating **C** and **D**, but not as low as **A**.

16 **D** From the formula sheet it can be seen that the decay constant (λ) is given by $\lambda = \frac{\ln 2}{t_{1/2}}$. Because the value of the decay constant is given in year^{-1} then the half-life ($t_{1/2}$) has units of years. Rearranging the equation gives $t_{1/2} = \frac{\ln 2}{\lambda} = \frac{\ln 2}{5.55 \times 10^{-10}}$. Substitution into the equation yields the correct value of 1.25×10^{9}.

17 **B** When thinking about which frame of reference is correct, you have to understand which frame is stationary and which is moving. As it is the muon that is being observed, the muon is moving in Earth's reference frame and stationary in its own. This would mean that from the muon's frame it is stationary and Earth is moving, therefore it is experiencing travelling for 626 m and being in existence for 2.1 μs. For it to exist at Earth's surface, time and distance have to change in Earth's reference frame.

18 **B** As both these moons orbit the common mass of Jupiter, from Kepler's Law it can be seen that $\frac{r_C^{3}}{T_C^{2}} = \frac{r_I^{3}}{T_I^{2}}$. Rearranging, this becomes $\frac{r_C^{3}}{r_I^{3}} = \frac{T_C^{2}}{T_I^{2}}$. Since $r_C = 4 \times r_I$ then $\frac{r_C^{3}}{r_I^{3}} = 64$. So $\frac{T_C^{2}}{T_I^{2}} = 64$, thus $\frac{T_C}{T_I} = 8$.

19 **B** The work done on the electron by the field is given by $W = qV$. This work results in an increase in the kinetic energy of the electron and so $qV = \frac{1}{2}mv^2$. Since q and m are unchanging, in order to generate a linear plot then axes should be V vs v^2.

20 **B** The proton has the smallest mass-to-charge ratio, and since there is a constant magnetic field and all velocities equal, will yield the smallest radius from the equation $r = \frac{mv}{qB}$, which is derived by equating the force experienced by a charged particle in a magnetic field with the centripetal force $\frac{mv^2}{r} = qvB$.

SECTION 2 PAGE 166

21 **a** The lemon will be moving slowest when it is at its maximum height. At this point the vertical velocity is 0. Since the horizontal velocity is constant then it must be moving faster at every other point in the trajectory, where a non-zero vertical component would be added to the horizontal component to determine total speed.

Criteria	Mark
Response correctly identifies slowest stage of trajectory and effectively justifies with reference to components of velocity.	2
Response identifies slowest stage of trajectory with no justification OR Attempts to analyse components of motion but is unable to identify slowest stage of trajectory	1

b The greatest speed will be when the vertical component is greatest (using logic above), which will occur when the lemon reaches the bottom of the cliff. So, data will be as follows:

$u = 40\,\text{m s}^{-1}$
$\theta = 20°$
$s = -70\,\text{m}$
$a = -9.8\,\text{m s}^{-2}$
$v_y^{2} = u_y^{2} + 2as$
$v_y^{2} = (40 \sin 20°)^2 + 2 \times -9.8 \times -70$
$\quad = 1559.164446$
$v_y = 39.48626 = 39\,\text{m s}^{-1}$
$v^2 = v_y^{2} + v_x^{2}$
Thus speed $= \sqrt{1559.164446 + (40 \cos 20°)^2}$
$\quad = 54.5237$
$\quad = 55\,\text{m s}^{-1}$

9780170449687

Criteria	Mark
Response substitutes correctly into correct equation with full working to arrive at correct answer.	3
Response substitutes correctly into correct equation with full working with an error or omission.	2
Response attempts to use equation of motion to analyse information from question.	1

22 a Same: Both particles must be of the same type of charge (negative) since both are deflected to the right as they enter the same field.

Different: The magnitude of the charge on the particles must be different because the different paths indicate that they are moving under the influence of forces of different magnitude. Because v and B are the same, a different force must result from a different charge: $F = qv_{\perp}B = qvB\sin\theta$. (Alternatively use the formula $r = \dfrac{mv}{qB}$

to justify the different radii: m, v and B are the same, so q must be different.)

Criteria	Mark
Response correctly relates cause and effect of the characteristic that is the same and the characteristic that is different.	4
Response correctly relates cause and effect of the characteristic that is the same and the characteristic that is different with some omission or error	3
Response successfully identifies the characteristic that is the same and the characteristic that is different.	2
Response identifies the characteristic that is the same OR the characteristic that is different.	1

b Both particles move in circular arcs. This occurs because charged particles in magnetic fields are subjected to a force perpendicular to their motion (right-hand rule). This defines the requirements for uniform circular motion, hence both particles move in circular arcs.

Criteria	Mark
Response successfully accounts for the circular arc of either or both particles.	2
Response identifies the circular arc OR Response attempts to account for the circular arc with some omission or error	1

23

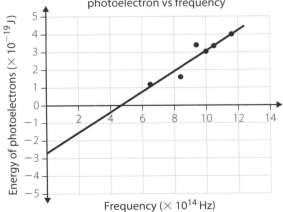

Maximum kinetic energy of photoelectron vs frequency

Work function is the negative y-intercept so would be approximately 2.8×10^{-19} J.

Threshold frequency is the x-intercept at approximately 4.7×10^{14} Hz.

Criteria	Marks
All points plotted accurately	6
Appropriate line of best fit is drawn and extrapolated to y-axis	
Value of work function correctly read as intercept of energy (vertical) axis	
Value of threshold frequency correctly read as intercept of frequency (horizontal) axis	
Most aspects of plotting, line of best fit, extrapolation and interpretation correct with some error(s) or omission(s)	4–5
Some aspects of plotting, line of best fit, extrapolation and interpretation correct with multiple errors or omissions	2–3
Some points plotted appropriately	1

24 a $F = \ell I_{\perp}B = \ell I\,B\sin\theta$
$= 0.35 \times 0.4 \times 25 \times 10^{-3} \times \sin 90°$
$= 3.5 \times 10^{-3}$ N down page

Criteria	Mark
Response successfully substitutes into appropriate equation to find solution and uses right-hand rule to determine direction.	2
Response finds solution but lacks direction OR Response attempts to solve with some errors	1

b When the conductor is turned through 90° there is no change in the angle that the conductor makes with the magnetic field lines. Consequently, the magnitude of the force is unchanged. (Note the direction of the force will now be to the left side of the page.)

Criteria	Mark
Response identifies that force will be unchanged in magnitude and explains successfully with statement regarding the unchanged value of angle in the equation.	2
Response identifies that force will be unchanged without explanation OR Response attempts to use flawed logic to explain correct response	1

25 a $F_c = \dfrac{mv^2}{r}$ is provided by $F_g = \dfrac{GMm}{r^2}$ so these equations can be equated:

$\dfrac{mv^2}{r} = \dfrac{GMm}{r^2}$

Dividing both sides by m and multiplying both sides by r gives

$v^2 = \dfrac{GM}{r}$

So $v = \sqrt{\dfrac{GM}{r}}$

Criteria	Mark
Response successfully equates F_c with F_G and simplifies to generate desired equation.	2
Response lacks derivation to arrive at remembered equation OR Response attempts to equate F_c with F_G with some errors	1

b $v = \sqrt{\dfrac{GM}{r}}$

Substituting into the equation above gives

$$v = \sqrt{\dfrac{6.67 \times 10^{-11} \times 6.0 \times 10^{24}}{(6370 + 1000) \times 10^3}}$$

$$7368.93 = 7.4 \times 10^3 \, \text{m s}^{-1}$$

Criteria	Mark
Response uses correct formula and working to arrive at correct answer with units.	2
Response features some relevant working.	1

26 a Sample answer

Main Sequence stars undergo fusion of hydrogen in their core. The process by which this occurs varies depending on the core temperature, mass and core concentrations of heavier elements. Lower mass stars such as Z undergo hydrogen fusion by means of the proton–proton chain. The sequence follows the steps outlined below and is severely limited by the rate of step 1.

1 $^1_1\text{H} + ^1_1\text{H} \rightarrow \ ^2_1\text{H} + ^0_1\text{e} + ^0_0\text{v}$

2 $^2_1\text{H} + ^1_1\text{H} \rightarrow \ ^3_2\text{He} + ^0_0\gamma$

3 $^3_2\text{He} + ^3_2\text{He} \rightarrow \ ^4_2\text{He} + ^1_1\text{H} + ^1_1\text{H}$

The net reaction releases 26.7 MeV and can be written as:

$4^1_1\text{H} \rightarrow \ ^4_2\text{He} + 2^0_1\text{e} + 2^0_0\text{v} + 2^0_0\gamma$

Larger mass stars such as Y with significant core concentrations of carbon undergo hydrogen fusion by means of the CNO cycle as outlined below.

1 $^{12}_6\text{C} + ^1_1\text{H} \rightarrow \ ^{13}_7\text{N} + ^0_0\gamma$

2 $^{13}_7\text{N} \rightarrow \ ^{13}_6\text{C} + ^0_1\text{e} + ^0_0\text{v}$

3 $^{13}_6\text{C} + ^1_1\text{H} \rightarrow \ ^{14}_7\text{N} + ^0_0\gamma$

4 $^{14}_7\text{N} + ^1_1\text{H} \rightarrow \ ^{15}_8\text{O} + ^0_0\gamma$

5 $^{15}_8\text{O} \rightarrow \ ^{15}_7\text{N} + ^0_1\text{e} + ^0_0\text{v}$

6 $^{15}_7\text{N} + ^1_1\text{H} \rightarrow \ ^4_2\text{He} + ^{12}_6\text{C}$

This process produces the same energy per helium nucleus created but occurs at a much higher rate.

Criteria	Marks
Correctly links p–p chain to low mass main sequence stars AND CNO cycle to high mass main sequence stars Indicates at least one difference AND similarity between the two processes	4
Correctly links p–p chain to low mass main sequence stars AND CNO cycle to high mass main sequence stars Indicates either a difference OR similarity between the two processes	3

Correctly links p–p chain to low mass main sequence stars AND CNO cycle to high mass main sequence stars	2
Response makes one correct statement about fusion reactions related to stimulus	1

b Sample answer

Fusion for Main Sequence stars such as X consists of the fusion of hydrogen to helium. This occurs a relatively low temperatures (10–15 million kelvin) when core hydrogen concentrations are high. Post-Main Sequence stars such as W fuse heavier elements in their core, such as carbon, oxygen, neon and silicon, with hydrogen fusion continuing in the outer shells. This requires much higher core temperatures (in excess of 100 million K) and higher concentrations of heavier elements.

Criteria	Marks
Correctly identifies hydrogen as a fuel for X and heavier elements as fuel for W stars AND Provides conditions under which these reactions take place	3
Correctly identifies hydrogen as a fuel for Main Sequence stars and heavier elements as fuel for Post-Main Sequence stars OR Provides conditions under which these reactions take place	2
Provides one correct piece of information related to fusion reaction of Main Sequence stars	1

c Sample answer

Medium mass stars such as V when on the Main Sequence are dense, yielding a high surface temperature but small surface area. As they progress from the Main Sequence, they increase in volume. This increase is driven by an increase in core temperature as they begin to fuse helium in their core. This is typically accompanied by a reduction in surface temperature and change in colour to become more red. As a result of the increased surface area, the luminosity of these stars will increase.

Criteria	Marks
Describes at least two evolutionary stages of a medium mass star Outlines internal processes that occur in these evolutionary stages Links these stages to changes in luminosity and surface temperature	3
Describes at least two evolutionary stages of a medium mass star Outlines internal processes that occur in these evolutionary stages	2
Describes a single evolutionary stage and the internal processes that occur.	1

9780170449687

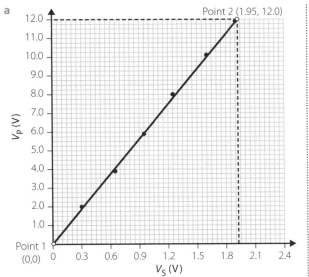

From the graph, the gradient $= \dfrac{12-0}{1.95-0} = 6.15$

Gradient $= \dfrac{\text{rise}}{\text{run}} = \dfrac{V_P}{V_S}$

Since $\dfrac{V_P}{V_S} = \dfrac{N_P}{N_S}$ (rewritten from formula sheet)

Then $\dfrac{1}{\text{gradient}} \times N_P = N_S$

Thus $N_S = \dfrac{1}{6.15} \times 500 = 81.25$ turns. The secondary coil has approximately 80 turns. Responses will vary depending on line of best fit drawn. Accept responses between 70 and 90 turns.

Criteria	Mark
Response features correctly plotted data, an appropriate line of best fit with two clearly indicated and well-separated points chosen on the line of best fit. Response uses these points to determine a value for the gradient of the line of best fit, which is then used to calculate N_S value.	4
Response features most of the following: correctly plotted data, an appropriate line of best fit with two clearly indicated and well-separated points chosen on the line of best fit, a value for the gradient of the line of best fit, which is then used to calculate N_S value.	3
Response uses line of best fit to determine a value for the gradient of the line of best fit, which is then used to calculate N_S value.	2
Response attempts an appropriate line of best fit.	1

b No sample answer is provided since it will depend on your plotting.

Criteria	Mark
Response reasonably relates the scatter of the points about the line of best fit with reliability	1

28 Similarities: the radius and speed are the same in each uniform circular motion situation, so the centripetal force will be constant throughout both circles and of the same magnitude given by the equation $F_c = \dfrac{mv^2}{r}$.

In both situations the centripetal force will be always directed towards the centre.

Differences: the centripetal force in the horizontal circle is entirely provided by the tension force, whereas in the vertical circle this is not the case. At the top of the vertical circle the force due to gravity provides some of the centripetal force.

The tension is constant in the horizontal circle whereas in the vertical circle the tension changes throughout the circle between a minimum at the top of the circle where $T = F_c - F_g$ and a maximum at the bottom where $T = F_c + F_g$.

Criteria	Mark
Response effectively compares clear, significant factors related to MOTION of disc in each situation including at least significant two similarities and differences.	4
Response compares clear, significant factors related to MOTION of disc with similarities and differences with an error or omission.	3
Response compares factors related to MOTION of disc and features similarities and/or differences, including some relevant information	2
Response notes a valid similarity and/or difference	1

29 Sample answer

Einstein identified some problems with the theory behind electromagnetism, which had some inconsistencies. To overcome these, he hypothesised that the speed of light was a constant in all inertial frames of reference and that in inertial frames of reference the laws of physics were the same. Due to constraints in technology at the time, Einstein was unable to carry out physical experiments to test his hypothesis, so he carried out 'thought' experiments such as the train traveller and stationary observer, who will see the speed of light at a constant speed and therefore time would pass at different rates for each. He was able to apply this model to several different situations that would happen in nature, such as length contraction, time dilation and relativistic momentum.

As the technology improved throughout the 20th century, the Theory of Special Relativity became testable and it became clear that Einstein's predictions were supported. The study of muons at Earth's surface showed that time dilation happened, as demonstrated by the extended lifespan of the muon. By being able to use particle accelerators, scientists have been able to observe and measure relativistic effects on particles showing relativistic momentum and mass. Because the evidence has supported the original hypothesis, there has been no need for a new hypothesis, which has led to the completion of the scientific method.

Criteria	Mark
Thorough evaluation of Einstein's thought experiments and how they followed the scientific method	5–6
Thorough evaluation of advances in technology to show how the Theory of Special Relativity has been supported	
Thorough understanding of the scientific method shown	
Comprehensive evaluation of Einstein's thought experiments and how they followed the scientific method	3–4
Comprehensive evaluation of advances in technology to show how the Theory of Special Relativity has been supported	
Comprehensive understanding of the scientific method shown	
Sound evaluation of Einstein's thought experiments and how they followed the scientific method	2
Sound evaluation of advances in technology to show how the Theory of Special Relativity has been supported	
Sound understanding of the scientific method shown	
Some relevant information.	1

30 a Sample answer

As can be seen from the equation, for both a controlled and an uncontrolled nuclear chain reaction to occur, an emitted neutron is required to initiate a subsequent fission of U-235 nuclei. In both reaction types, a critical mass of fissile fuel must be present to enable enough collisions between neutrons and the nucleus to occur AND a moderator (heavy water or graphite) must be used to slow the fission product neutrons for this collision to be successful. Note that the reaction produced three neutrons. For a controlled reaction to take place, exactly one of these produced neutrons needs to be captured by a subsequent U-235 atom. Control rods made of cadmium are able to absorb neutrons so that an average of exactly one subsequent fission takes place per fission reaction. In contrast, if any more than one of these neutrons is captured by subsequent U-235 atoms, an uncontrolled reaction will proceed.

Criteria	Marks
Identifies the role neutrons play in the initiation of these reactions	3
Correctly identifies the differing quantitative requirements for subsequent reactions caused by emitted neutrons in both types and neutron capture in controlled reactions and how this can be achieved	
Identifies the role neutrons play in the initiation of these reactions	2
Correctly identifies the differing quantitative requirements for subsequent reactions caused by emitted neutrons for each of these types of reactions	
Identifies the role neutrons play in the initiation of these reactions	1

b Sample answer

$(2.05 \times 10^8)(2.563 \times 10^{21})(15.45) = 8.117\,661\,75 \times 10^{30}\,\text{eV}$
Multiplying by 1.602×10^{-19} gives $1.300\,449\,412 \times 10^{12}$
$= 1.30 \times 10^{12}\,\text{J}$

Criteria	Marks
Correctly calculates the energy released	2
Correctly converts to joules	
Correct significant figures (3)	
Correctly calculates the energy released with either incorrect significant figures or units (eV)	1

c i Sample answer

$$^{238}_{92}\text{U} \rightarrow \,^{234}_{90}\text{Th} + \,^{4}_{2}\text{He}$$

Criteria	Marks
Correctly writes balanced nuclear equation	2
Writes a nuclear equation that features an error or omission	1

ii Sample answer

Nuclei heavier than lead ($Z = 82$) are inherently unstable due to the increased magnitude of the repulsive force acting between all protons in the nucleus that overcomes the short-range strong nuclear force that acts only between neighbouring nucleons. Alpha decay is a way of decreasing this repulsive force; the proton:neutron ratio is improved and this change increases stability.

Criteria	Marks
Identifies the major reason for instability of U-238 as nuclear size	2
Links the emission of an alpha particle with the reduction in nuclear size	
Identifies the major reason for instability of U-238 as nuclear size or identifies that alpha emission reduces nuclear size	1

31 Accepted points of comparison.

	Qualitative	Quantitative
Similarity	$v_x = $ constant	$v_x = 4.32 \times 10^6\,\text{m s}^{-1}$
	v_y changes due to constant acceleration	
	Motion is parabolic arc/projectile	
Difference	Arc curves down in gravitational field	$a_E = 9.67 \times 10^{15}\,\text{m s}^{-2}$
	Arc curves up in electric field	$a_g = 9.8\,\text{m s}^{-2}$
	Arc is flatter in gravitational field	with relevant calculations included
	Arc is sharper in electric field	velocity at any point in time is greater, with relevant calculations included
	Motion towards arrows in gravitational field	
	Motion opposite to arrows in electric field	
	Acceleration is greater in electric field than gravitational field	

9780170449687

Criteria	Mark
Response effectively compares clear, significant factors related to motion of electrons in each situation, including at least two similarities and differences and including both qualitative and quantitative information, with calculations or processes to correctly determine relevant values (a_g, a_E, v_x, v_y) for both situations.	7
Response compares clear, significant factors related to motion with similarities and differences, including both qualitative and quantitative information, with calculations or processes to determine a less relevant value (F, K) for both situations.	5–6
Response compares factors related to motion and features similarities and/or differences, including some qualitative and/or quantitative information OR Response provides factors related to motion, including some qualitative and/or quantitative information	3–4
Response notes a valid similarity and/or difference OR Response provides a factor related to motion, including some qualitative or quantitative information	1–2

32 Suggested response would include the following points.

Cause:

Phased AC current is provided to opposite pairs of stator coils (seen in diagram).

Thus, change in flux experienced by the bars of the squirrel cage rotor (seen in diagram).

Therefore, emf is induced in the squirrel cage rotor (Faraday's Law).

Thus, current flows in circuit along bars and connecting end plates (seen in diagram).

The direction of this current establishes a magnetic field that interacts with the stator coil field.

Effect:

This interaction results in a magnetic force

Which causes a torque on the squirrel cage rotor.

OR

Effect:

This current in the squirrel case rotor means the squirrel cage rotor experiences a motor effect force.

Which causes a torque on the squirrel cage rotor.

Criteria	Mark
A cause and effect response effectively incorporates aspects of the diagram to follow a chain of logic that links the AC supply current through all key steps to the delivered torque	5
A cause and effect response links the AC supply current through some key steps to the delivered torque with a significant error or omission or without reference to diagram	4

Criteria	Mark
Response outlines a substantially correct process with some significant errors or omissions	3
Response outlines some relevant processes for functioning of AC induction motor with various key omissions	2
A limited outline features an aspect of AC induction motor process	1

33 Sample answer

The question states that *It is said that 'if sufficient evidence supports a hypothesis then it can be considered a valid theory'*.

After Max Planck proposed that light was quantised and that the energy of each photon could be determined using the equation $E = hf$, Einstein took this further by proposing that if each photon had energy exceeding a specific energy level then it would be able to remove or 'knock out' electrons from a metal, behaving much like a billiard ball. He proposed that this would explain the photoelectric effect observed experimentally by Philipp Lenard. Using conservation of energy, Einstein showed that the maximum kinetic energy of the ejected photoelectron could be found by using the photoelectric effect formula $K_{max} = hf - \phi$ with ϕ being the work function unique to the metal.

The photoelectric effect was tested by Millikan when he improved the original experiment to the point where it was accurate enough to be able to test and support Einstein's theory. The experiment involved shining incidental light onto a metal surface that was in a vacuum chamber. There was an anode across from the metal surface and a potential difference was placed across these. This allowed experimental evidence to be obtained to support the photoelectric effect theory.

The implication of this was to show that light behaved as both a wave and a particle in certain situations. Consequently, this evidence can be considered to be sufficient to consider Einstein's proposed theory valid.

De Broglie proposed that if light can behave as both a wave and a particle then matter should also be able to do so. He based this on the understanding that he could quantitively describe the wave–particle duality by using Einstein's energy–mass equivalence formula and Planck's hypothesis, showing $\lambda = \dfrac{h}{mc}$. de Broglie theorised that if you replaced c with v then it should apply to all substances and not just electromagnetic radiation.

Davison and Germer provided experimental evidence to support this theory by firing electrons at nickel crystals to show that as the electrons passed through the gaps in the structure they diffracted and caused an interference pattern to be formed, thus supporting the theory that electrons behaved as both particles and waves.

Consequently, this evidence can be considered to be sufficient to consider de Broglie's proposed theory valid.

This had the effect of explaining why electrons when in orbit around a nucleus, as in Bohr's atomic model, do not emit electromagnetic radiation, as they are standing waves and therefore are not accelerating.

Criteria	Marks
Effective judgement made based upon clearly stated criteria with clear and logical reasoning	8–9
Extensive understanding of the evidence for the wave nature of light	
Extensive understanding of the evidence for the particle nature of light	
Extensive understanding of de Broglie's matter wave theory	
Extensive understanding of Einstein's wave–particle theory	
Description of implications of each theory	
Thorough understanding of the evidence for the wave nature of light	5–7
Thorough understanding of the evidence for the particle nature of light	
Thorough understanding of de Broglie's matter wave theory	
Thorough understanding of Einstein's wave–particle theory	
Judgement made based on logical reasoning	
Description of implications of each theory	
Sound understanding of the evidence for the wave nature of light	3–4
Sound understanding of the evidence for the particle nature of light	
Sound understanding of de Broglie's matter wave theory	
Sound understanding of Einstein's wave–particle theory	
Sound reasoning	
Some understanding of the evidence for the wave nature of light	2
Some understanding of the evidence for the particle nature of light	
OR	
Some understanding of de Broglie's matter wave theory	
Some understanding of Einstein's wave–particle theory	
Some understanding of the evidence for the wave nature of light	1
OR	
Some understanding of the evidence for the particle nature of light	
OR	
Some reasoning in answering the statement	

9780170449687